Acknowledgments

Thank you to everyone who helped put *Geometry Made Easy, Common Core Edition* together. That includes Kimberly Knisell, Director of Math and Science in the Hyde Park, NY School District. Marin Malgieri who contributed proof reading skills and assistance with several topics. Jennifer Criser-Eighmy who proof read the grammar and punctuation. Allen Miller and Nancy Brush for their proofreading skills. Julieen Kane, the graphic designer at Topical Review Book Company did a fabulous job with the diagrams and drawings. Lastly, but definitely very importantly, Keith Williams, the owner of Topical Review Book Company deserves a major thank you. We have worked together for almost 20 years now and it is always a pleasure. His company provides excellent materials for students to use at a reasonable price. I used the "Little Green Regents Review Books" way back in the old days when I was in high school. The company was already 25 years old at that time. I am honored to have my work published by Topical Review Book Company and to share in their history.

Introduction

As the Common Core Standards are implemented nationally, there are new methods of teaching material that was taught in a more traditional way in the past. The new presentations associated with the standards will help our students to become "college and career ready." *Geometry Made Easy, Common Core Edition*, is meant to be a reference guide for the mathematical procedures needed to help the student complete their work in using the standards. It is not meant to be a curriculum guide and is not designed to replace any teaching methods that are used in the classroom. It is my hope that this student friendly handbook will help each geometry student to achieve success in completing his/her study of the Geometry Common Core Standards.

Sincerely,

MaryAnn Casey,
B.S. Mathematics, M.S. Education

GEOMETRY MADE EASY
Common Core Standards Edition

Table of Contents

UNIT 1: FOUNDATIONS 1
  1.1 Applying Geometric Concepts 2
  1.2 Geometric Solutions Using Proofs 4
  1.3 Geometric Terms..13

UNIT 2: CONGRUENCE, PROOFS AND CONSTRUCTIONS ... 17
  2.1 Transformational Geometry Terms....................18
  2.2 Rigid Motions 20
  2.3 Geometric Terms..32
  2.4 Proving Geometric Theorems..........................36
  2.5 Constructing Lines And Angles.......................46
  2.6 Inscribing Polygons in a Circle 54

UNIT 3: SIMILARITY PROOFS, AND TRIGONOMETRY 57
  3.1 Transformation and similarity 58
  3.2 Similarity and Congruence of Triangles...........64
  3.3 Similarity and Congruence of Polygons...........83
  3.4 Right Triangles and Trigonometric Ratios....................93

UNIT 4: EXTENDING TO THREE DIMENSIONS 107
  4.1 Volume..108
  4.2 3-Dimensional Figures and Their Properties...............111
  4.3 Changing a 2-Dimensional Figure
      To A 3-Dimensioal Figure 115

UNIT 5: CONNECTING ALGEBRA AND GEOMETRY
  THROUGH COORDINATES 117
  5.1 Graphing Basics..118
  5.2 Coordinate or Analytic Proof Example 127
  5.3 Partitioning a Segment in a Given Ratio 129
  5.4 Perimeter and Area 132
  5.5 Parabolas...135

Geometry Made Easy Handbook
Common Core Standards Edition

By:
Mary Ann Casey
B. S. Mathematics, M. S. Education

GEOMETRY MADE EASY
Common Core Standards Edition

Table of Contents

UNIT 6: CIRCLES WITH AND WITHOUT COORDINATES .. 139
    6.1 Circumference and Area of a Circle 140
    6.2 Circles and Angles 144
    6.3 Similarity of Circles 147
    6.4 Circles and Their Angles and Arcs 149
    6.5 Angles of Sectors 153
    6.6 Circles and Segments 160
    6.7 Equation of a Circle 167
    6.8 Circles and Polygons 169

CORRELATIONS TO CCSS 172

INDEX 174

FOUNDATIONS

- Apply geometric concepts.

- Solve using proofs.

- Recognize and use geometric terms.

1.1

APPLYING GEOMETRIC CONCEPTS

Modeling or modeling situations are terms that are used in the Common Core Standards to describe types of problems that involve word problems, applications, real world problems, and story problems. They set up a scenario and the student is asked to solve a problem, answer a question, or to determine the best solution. Naturally there are often many ways to present a solution. Typically the student is asked to support or justify the solution. The justification or support must be done using logical reasoning and standard mathematical procedures. The justification can be a description of the procedures used (See Example 1), or it can be a written algebraic explanation (see Example 2). These are already familiar. Another type of justification or support is a geometric proof. Examples of geometric proofs are on the pages that follow.

ALGEBRAIC SOLUTIONS

Steps

1) Read the problem carefully.

2) Decide what, if any, formulas are needed. Make a diagram if possible.

3) Write an equation or substitute in the appropriate formula.

4) Solve.

5) Answer the question completely with a written conclusion.

Examples

❶ A parking garage is being constructed to contain a maximum of 100 cars. The space for each car is in the shape of a rectangle. If the average car requires a parking space that is 6 feet by 12 feet, how many square feet of parking space must be constructed? Justify your answer.

Car:

$A = lw$

$A = (6)(12)$

$A = 72 \ sq \ ft$

Garage:

$100(72) = 7200 \ sq \ ft$

Conclusion: Using the formula for the area of a rectangle, each car requires 72 square feet of space. Since there are a maximum of 100 cars, the garage must contain at least 7,200 square feet of parking space.

Geometry Made Easy – Common Core Standards Edition

❷ Tom and Jerry are making a vegetable garden in their yard. They want to be creative, so they are designing it in the shape of a parallelogram instead of the usual rectangle or square. They want to make one side of the garden twice as long as the other. They have 24 yards of fencing to enclose the garden. What are the dimensions of the garden in feet? Would there be a difference in the measurements if the garden is rectangular instead of being a parallelogram in shape? Why or why not? Would the areas of the parallelogram shaped garden and the rectangular garden be the same? Explain.

Note: Pay attention to yards vs feet.

Let x = width of the garden in yards
$2x$ = length of the garden
Perimeter = $2l + 2w$
$24 = 2(x) + 2(2x)$
$24 = 6x$
$x = 4$ yards, $2x = 8$ yards
1 yd = 3 ft
$x = 12$ ft
$2x = 24$ ft

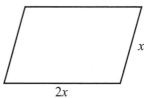

Conclusion: The garden has one pair of opposite sides that are 12 feet in length, and the other two sides are 24 feet in length. If the garden is made in the shape of a rectangle, the dimensions would still be 12 × 24 feet since the formulas for the perimeter of a rectangle and the perimeter of a parallelogram are the same. The area of the rectangular garden would be more. The rectangular garden would have an area equal to the product of the length and width. The parallelogram area would be the product of the length and the altitude of the parallelogram. The altitude is the perpendicular distance between the two lengths, and is shorter than the "width" measurement of the parallelogram. The area of the parallelogram is less than the area of the rectangular garden.

GEOMETRIC SOLUTIONS USING PROOFS

A geometric proof is a specific kind of presentation of support for the multiple logical steps used to solve a geometry problem. The problem may be presented as a modeling situation, although often it is simply a problem presented directly in its mathematical form. The examples that follow are shown in mathematical form. They do not relate directly to a modeling situation, but they demonstrate the use of proofs in geometric problems. Problems presented in modeling form can be interpreted mathematically and then solved using a proof.

Steps

1) Read the problem. Note what is given and what needs to be proven.

2) Make a diagram and label it.

3) Consider the types of proofs and choose the format that works best for the problem. Examples are given below.

4) Start with step 1 by stating the information given.

5) Continue with logical steps leading to the statement that is to be proven. Each step must be based upon information given in the problem, or already established in the proof.

6) The final step of a proof is the statement that is to be proven.

Key Idea: Each step in a geometric proof of any kind must be based either on information given in the problem, or on steps previously completed.

TYPES OF PROOFS

- Euclidean Proof (also called Statement-Reason or Two Column Proof)
- Paragraph Proof
- Flow Proof or Flowchart
- Analytic or Coordinate Proof
- Proof by Rigid Motion
- Indirect Proof (also called Proof by Contradiction)

In the proofs throughout this handbook, abbreviations of commonly used geometric statements are used. Although these are widely accepted, not all teachers accept abbreviations. Follow your teacher's directives to receive full credit for your work.

Geometry Made Easy – Common Core Standards Edition

EUCLIDEAN GEOMETRY PROOF

A **Euclidean Geometry Proof** is a formal "statement/reason" or "2-column"proof. Each step in the progress toward the conclusion is considered to be a "statement" and is written down. Next to the statement, the mathematical reason that allowed the step to be done is written. <u>Each step in a proof must be based on steps already completed or on given information.</u> The last step, or statement, will be the conclusion required and next to it, the final "reason" used to get to that conclusion.

Example This problem was chosen to demonstrate the methods used to solve problems using the formal proof process. Check with your teacher for specific instructions, as there are many ways to do a proof.

❶ **Given:** $\triangle ABC$
   Prove: $m\angle 1 + m\angle 2 + m\angle 3 = 180$

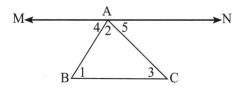

Statement	Reason
1. $\triangle ABC$	1. Given.
2. Through point A, draw line $\overline{NM} \parallel \overline{BC}$ (Label angles as shown.)	2. Through a point not on a given line, there exists one and only one line parallel to the given line.
3. $\angle 1 \cong \angle 4$, $\angle 3 \cong \angle 5$	3. If two parallel lines are cut by a transversal, alternate interior angles are $\cong$.
4. $m\angle 1 = m\angle 4$; $m\angle 3 = m\angle 5$	4. Definition of $\cong$ angles.
5. $\angle 4$ and $\angle BAN$ are supplementary.	5. Two angles that form a straight line are supplementary.
6. $m\angle 4 + m\angle BAN = 180$	6. Definition of supplementary angles.
7. $m\angle BAN = m\angle 2 + m\angle 5$	7. The whole is equal to the sum of the parts.
8. $m\angle 4 + m\angle 2 + m\angle 5 = 180$	8. Substitution.
9. $m\angle 1 + m\angle 2 + m\angle 3 = 180$	9. Substitution.

1.2

❹ Given: Parallelogram *NEMO*,
diagonal *NCDM*,
$\overline{OC} \perp \overline{NM}$, $\overline{ED} \perp \overline{NM}$

Prove: $\triangle OCM \cong \triangle EDN$

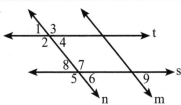

Statement	Reason
1. *NEMO* is a parallelogram $\overline{OC} \perp \overline{NM}$, $\overline{ED} \perp \overline{NM}$	1. Given.
2. $\overline{OM} \parallel \overline{NE}$, $\overline{OM} \cong \overline{NE}$	2. Opposite sides of a parallelogram are parallel and congruent.
3. $\angle OCM$ and $\angle EDN$ are right angles.	3. Perpendicular lines form right angles.
4. $\angle OCM \cong \angle EDN$	4. All right angles are congruent.
5. $\angle NMO \cong \angle MNE$	5. When two parallel lines are cut by a transversal, alternate interior angles are congruent.
6. $\triangle OCM \cong \triangle EDN$	6. AAS ≅ AAS

❺ Parallel Lines Proof
Given: $n \parallel m$, $t \parallel s$
Prove: $\angle 1 \cong \angle 9$

Statement	Reason
1. $n \parallel m$, $t \parallel s$	1. Given.
2. $\angle 1 \cong \angle 8$	2. If two parallel lines are cut by a transversal, corresponding angles are congruent. (*t* and *s* are the parallel lines, *n* is the transversal.)
3. $\angle 8 \cong \angle 6$	3. Vertical angles are congruent.
4. $\angle 1 \cong \angle 6$	4. Transitive Property (or Substitution).
5. $\angle 6 \cong \angle 9$	5. If two parallel lines are cut by a transversal, corresponding angles are congruent. (*n* and *m* are the parallel lines, *s* is the transversal.)
6. $\angle 1 \cong \angle 9$	6. Transitive Property (or Substitution).

Geometry Made Easy – Common Core Standards Edition

PARAGRAPH PROOF

A paragraph proof is also called an "informal proof." A plan is made and the statements and reasons are written in the form of a paragraph. Included must be: the given information; what is to be proven; a description of the deductive reasoning steps; the reasons being used; a diagram when possible; and a conclusion. It could be thought of as writing a formal two column proof in a more conversational form, but all the information must be included.

Example **Paragraph Proof**

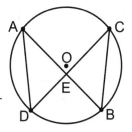

Given: Chords $\overline{AB}$ and $\overline{CD}$ of circle O intersect at E, an interior point of circle O; chords $\overline{AD}$ and $\overline{CB}$ are drawn.

Prove: $(AE)(EB) = (CE)(ED)$

We are given chords $\overline{AB}$ and $\overline{CD}$ of circle O. They intersect at E, an interior point of circle O. Chords $\overline{AD}$ and $\overline{BC}$ are drawn. $\angle A \cong \angle C$ because inscribed angles of a circle that intercept (subtend) the same arc are congruent. $\angle AED \cong \angle CEB$ because they are vertical angles. $\triangle AED$ is similar to $\triangle CEB$ because when two angles of one triangle are congruent to two angles of another triangle, the triangles are similar. In similar triangles, corresponding sides are proportional, so $\dfrac{AE}{CE} = \dfrac{ED}{EB}$. Therefore $(AE)(EB) = (CE)(ED)$ because in a proportion, the product of the means equals the product of the extremes.

1.2

FLOW PROOF OR FLOWCHART

In a flow or chart proof, each statement is written in a box and the reason it is used is written under the box. The boxes are connected with arrows to show the sequence of the proof in reaching the conclusion. Again, the given information, what is to be proven, and a diagram are parts of this type of proof.

Examples Flow Proofs

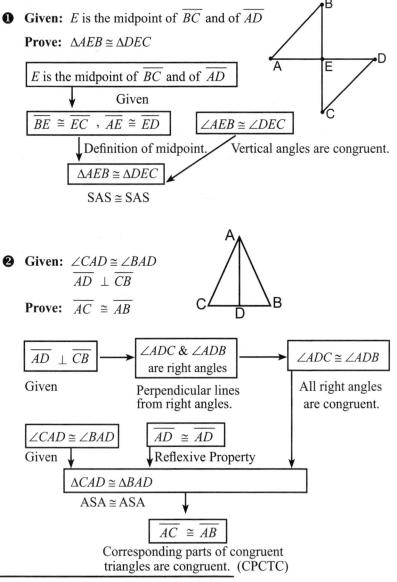

❶ **Given:** E is the midpoint of $\overline{BC}$ and of $\overline{AD}$

Prove: $\triangle AEB \cong \triangle DEC$

E is the midpoint of $\overline{BC}$ and of $\overline{AD}$

Given

$\overline{BE} \cong \overline{EC}$, $\overline{AE} \cong \overline{ED}$

Definition of midpoint.

$\angle AEB \cong \angle DEC$

Vertical angles are congruent.

$\triangle AEB \cong \triangle DEC$

SAS ≅ SAS

❷ **Given:** $\angle CAD \cong \angle BAD$

$\overline{AD} \perp \overline{CB}$

Prove: $\overline{AC} \cong \overline{AB}$

$\overline{AD} \perp \overline{CB}$

Given

$\angle ADC$ & $\angle ADB$ are right angles

Perpendicular lines from right angles.

$\angle ADC \cong \angle ADB$

All right angles are congruent.

$\angle CAD \cong \angle BAD$

Given

$\overline{AD} \cong \overline{AD}$

Reflexive Property

$\triangle CAD \cong \triangle BAD$

ASA ≅ ASA

$\overline{AC} \cong \overline{AB}$

Corresponding parts of congruent triangles are congruent. (CPCTC)

Geometry Made Easy – Common Core Standards Edition

ANALYTIC OR COORDINATE PROOF

A coordinate proof is used when the problem can be represented on a coordinate plane. The points involved have (x, y) coordinates. The proof usually requires the use of the distance formula, the slope formula, the midpoint formula, or some combination of the three.

Example

Given: $\triangle ABC$ with vertices $A(7, 2)$, $B(4, 4)$, and $C(2, 1)$
Prove: $\triangle ABC$ is an isosceles right triangle.

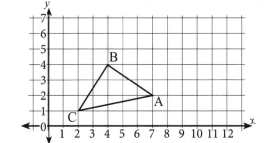

Plan: There are two equally appropriate methods that can be used here.

1) **Prove:** $\overline{AB} \cong \overline{BC}$ and $\overline{AB} \perp \overline{BC}$.

Formulas needed are distance and slope.

$$d = \sqrt{(x_2 - x_1)^2 + (y_2 - y_1)^2} \qquad m = \frac{y_2 - y_1}{x_2 - x_1}$$

$$d_{AB} = \sqrt{(7-4)^2 + (2-4)^2} = \sqrt{9+4} = \sqrt{13} \qquad m_{AB} = \frac{2-4}{7-4} = \frac{-2}{3} = -\frac{2}{3}$$

$$d_{BC} = \sqrt{(4-2)^2 + (4-1)^2} = \sqrt{4+9} = \sqrt{13} \qquad m_{BC} = \frac{4-1}{4-2} = \frac{3}{2}$$

Conclusion: $\triangle ABC$ is a right isosceles triangle because its two legs have the same length, and they are perpendicular to each other since their slopes are negative reciprocals of each other.

2) **Prove:** $\overline{AB} \cong \overline{BC}$ and $AC^2 = AB^2 + BC^2$.

Formulas needed are distance and Pythagorean Theorem.

$$d = \sqrt{(x_2 - x_1)^2 + (y_2 - y_1)^2} \qquad c^2 = a^2 + b^2$$

$$d_{AB} = \sqrt{(7-4)^2 + (2-4)^2} = \sqrt{9+4} = \sqrt{13} \qquad \left(\sqrt{26}\right)^2 = \left(\sqrt{13}\right)^2 + \left(\sqrt{13}\right)^2$$

$$d_{BC} = \sqrt{(4-2)^2 + (4-1)^2} = \sqrt{4+9} = \sqrt{13} \qquad 26 = 13 + 13$$

$$d_{AC} = \sqrt{(7-2)^2 + (2-1)^2} = \sqrt{25+1} = \sqrt{26} \qquad 26 = 26$$

Conclusion: $\triangle ABC$ is an isosceles triangle because its two legs are the same length. It is a right triangle because the square of the hypotenuse is equal to the sum the of the squares of the two legs.

PROOF USING RIGID MOTION

Rigid motion transformations are reflections, rotations, and translations. If it can be shown that the figure presented in the problem shows one or more of these transformations, a rigid motion proof can be used. Any acceptable style of proof can be used. (See Unit 2 for information about rigid motions.)

PARAGRAPH PROOF USING RIGID MOTION

Example

Given: $\triangle ABD$ is isosceles. $\triangle ABD \xrightarrow{\; r\;\overline{BD}\;} \triangle CBD$, quadrilateral $ABCD$ is formed.

Prove: $ABCD$ is a rhombus.

$\overline{AB} \cong \overline{AD}$ because an isosceles triangle has two congruent sides.

A reflection is a rigid motion. Therefore distance is preserved, making $\overline{AB} \cong \overline{CB}$, $\overline{AD} \cong \overline{CD}$. Using substitution we can say that $\overline{AB} \cong \overline{CD}$, $\overline{AD} \cong \overline{CB}$. $ABCD$ is a parallelogram because if a quadrilateral has both pair of opposite sides that are congruent, it is a parallelogram. Since $\overline{AB} \cong \overline{CB}$, $\overline{AD} \cong \overline{CD}$, $ABCD$ is a rhombus because a parallelogram with adjacent sides that are congruent is a rhombus.

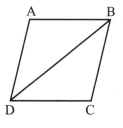

INDIRECT PROOF

Indirect proof is also called "Proof by Contradiction." This proof requires that we assume the conclusion to be drawn is false! The use of the false conclusion is called an "assumption" and is an important part of this type of proof. Through the progress of the proof, logical reasoning leads to a contradiction of the hypothesis (the "given") or some other known fact such as a theorem, definition, postulate, etc. By reaching an incorrect conclusion based on the assumption (the false original conclusion), we prove the original conclusion must be true.

An indirect proof is often used to prove that something is *not* true.

Steps

1) Write the given.

2) Using the "prove" statement, assume it is *not* true.

3) Proceed as usual trying to prove the assumption in step 2.

4) A contradiction will appear in the proof
   -- a statement that is opposite something that is given or known.

5) The contradiction allows the conclusion that the original
   "prove" statement must be true.

Note: An indirect proof can be done using any of the acceptable
   proof styles.

Examples

❶ **Paragraph Proof By Contradiction:**

Mary lives on Avenue M at the corner labeled 3. Nancy lives on Avenue N at the corner labeled 1. Avenue R cuts across both avenues. The city map shows that avenues M and N are NOT parallel. Nancy says the corner angles 1 and 3 are not equal. Write a geometric proof to explain.

Given: M is not $\parallel N$

Prove: $\angle 1$ is not $\cong \angle 3$

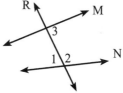

We are given that line M is not parallel to line N. Assume that $\angle 1 \cong \angle 3$. $\angle 1$ and $\angle 3$ are alternate interior angles by definition. This makes $M \parallel N$ because if two lines are cut by a transversal and alternate interior angles are congruent, the lines are parallel. This conclusion is contradictory to the given statement. Therefore if $\angle 1$ is not congruent to $\angle 3$, then M is not parallel to N.

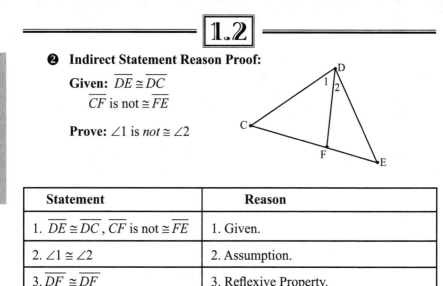

❷ **Indirect Statement Reason Proof:**

Given: $\overline{DE} \cong \overline{DC}$
$\overline{CF}$ is not $\cong \overline{FE}$

Prove: $\angle 1$ is *not* $\cong \angle 2$

Statement	Reason
1. $\overline{DE} \cong \overline{DC}$, $\overline{CF}$ is not $\cong \overline{FE}$	1. Given.
2. $\angle 1 \cong \angle 2$	2. Assumption.
3. $\overline{DF} \cong \overline{DF}$	3. Reflexive Property.
4. $\triangle CDF \cong \triangle EDF$	4. SAS $\cong$ SAS
5. $\overline{CF} \cong \overline{FE}$	5. CPCTC *
6. $\angle 1$ is not $\cong \angle 2$	6. Contradiction of step 5 and given.

* Common abbreviation for Corresponding Parts of Congruent Triangles are Congruent.

GEOMETRIC TERMS

COMMON GEOMETRIC TERMS, DEFINITIONS, AND SYMBOLS

The tables that follow contain informal summaries and "short cuts" for many of the symbols, terms, and definitions that are used in geometry. Some can be used as reasons in a proof and others are descriptive terms. For more formal definitions, use your text or a math dictionary. Always listen to the directives of your teacher in using symbols or short cuts.

SYMBOLS USED IN PROBLEMS AND PROOFS

Description	Symbol	Example
Parallel	$\parallel$	$m \parallel n$
Perpendicular	$\perp$	$m \perp n$
Congruent	$\cong$	$\angle A \cong \angle B$
Approximately equal to	$\approx$	$\sqrt{3} \approx 1.73$
Similar	$\sim$	$\triangle ABC \sim \triangle DEF$
Maps to	$\xrightarrow{R_{45°}}$ or $\xrightarrow{R_{45°}}$	$\triangle ABC \xrightarrow{R_{45°}} \triangle A'B'C'$ or $\triangle ABC \xrightarrow{R_{45°}} \triangle A'B'C'$
Circle	$\odot$	$\odot P$

CONGRUENT VS SIMILAR

Congruent: Geometric figures are congruent if they have corresponding sides that are equal in measure and corresponding angles that are equal in measure. Rigid motion transformations result in figures that are congruent.

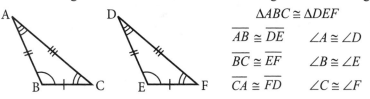

$$\triangle ABC \cong \triangle DEF$$

$\overline{AB} \cong \overline{DE}$	$\angle A \cong \angle D$
$\overline{BC} \cong \overline{EF}$	$\angle B \cong \angle E$
$\overline{CA} \cong \overline{FD}$	$\angle C \cong \angle F$

Similar: Geometric figures are similar if they have corresponding angles that are equal in measure and corresponding sides that are proportional. The ratio of the proportion of the corresponding sides is called the constant of proportionality. Transformations involving dilations result in similar figures.

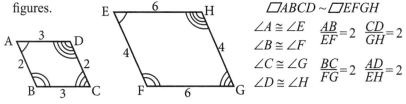

$$\square ABCD \sim \square EFGH$$

$\angle A \cong \angle E \quad \dfrac{AB}{EF} = 2 \quad \dfrac{CD}{GH} = 2$

$\angle B \cong \angle F$

$\angle C \cong \angle G \quad \dfrac{BC}{FG} = 2 \quad \dfrac{AD}{EH} = 2$

$\angle D \cong \angle H$

COMMONLY USED TERMS AND DEFINITIONS

Word, Term, and Mathematical Symbol	Diagram and Labels	Brief Definition
Point	• P	A location in space. Has no length, width or depth.
Line $\overleftrightarrow{AB}$		Has infinite length, no width.
Plane	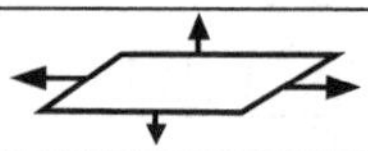	Has infinite length and width. No depth.
Segment $\overline{AB}$		Part of line between 2 points.
Collinear Points ABC	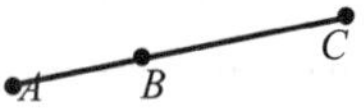	Points that are on the same line /line segment.
Ray $\overrightarrow{CD}$		A partial line that starts at a point & goes in one direction.
Angle $\angle ABC$ or $\angle ABC$	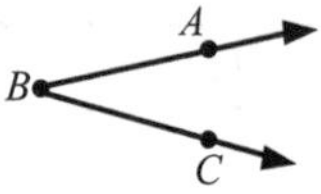	Formed when 2 rays meet at a point or when 2 lines intersect. Label with the vertex (point) at the center. The size of the opening between the rays is measured in degrees or radians.
Bisect line bisector-Figure 1 angle bisector-Fig. 2	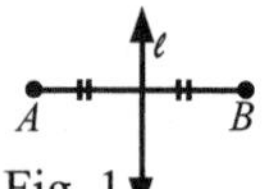 Fig. 1 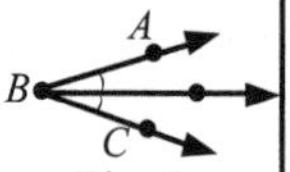 Fig. 2	To cut in half: A bisector cuts a line segment into 2 ≅ parts, an angle into 2 ≅ angles.
Perpendicular $\overleftrightarrow{CD} \perp \overleftrightarrow{AB}$	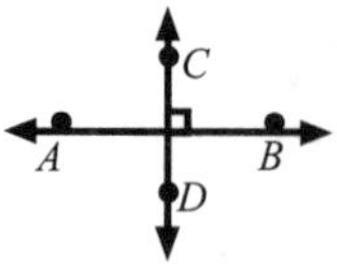	2 lines that intersect at right angles.
Parallel $\overleftrightarrow{CD} \parallel \overleftrightarrow{AB}$	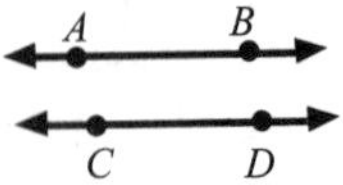	2 lines in a plane that never meet.
Right Angle	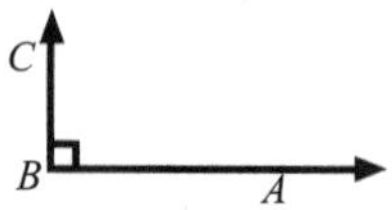	Measures 90°.

Geometric Terms (side tab)

Geometry Made Easy – Common Core Standards Edition

1.3

Geometric Terms

Term		Description
Straight Angle		Measures 180°.
Acute Angle		Measures more than 0°, less than 90°.
Obtuse Angle		Measures more than 90°, less than 180°.
Reflex Angle		Measures more than 180°, less than 360°.
Equiangular		All angles are congruent.
Equilateral		All sides are congruent.
Scalene		All sides of a figure are different lengths.
Isosceles $\triangle$		2 sides are congruent: $\triangle ABD$ is isosceles.
vertex angle and base angles in an isosceles $\triangle$		$\overline{AD} \cong \overline{DB}$ and $\overline{AB}$ is the base. D is the vertex angle, A and B are the base angles
Regular Polygon		A polygon with equal sides & equal angles. 5, 6, 8, 10 and 12 sided polygons are often used.
Adjacent Angles $\angle 1$ and $\angle 2$ are adjacent $\angle$s.		Next to each other. Angles which share only one side and a vertex but have no interior points in common.
Linear Pair		2 adjacent angles formed by the intersection of 2 lines. They are supplementary angles.
Opposite Angles $\angle 2$ and $\angle 4$ are opposite.		Across from each other. Not sharing a side or a vertex.
Vertex		The "point" of an angle, the corner of a polygon.
Diagonal		Connects 2 opposite vertices (corners) in a geometric figure.
Consecutive Angles /Sides $\angle A$ and $\angle B$ $\quad \angle C$ and $\angle D$ $\angle B$ and $\angle C$ $\quad \angle D$ and $\angle A$ $\overline{AD}$ and $\overline{DC}$ $\quad \overline{BC}$ and $\overline{AB}$ $\overline{DC}$ and $\overline{CB}$ $\quad \overline{AB}$ and $\overline{DA}$		Sides or angles that are "one after the other".

Geometry Made Easy – Common Core Standards Edition

Pi π	π is an irrational number. The symbol π should be included in your answer if the problem says leave in terms of Pi .	The ratio of circumference to the diameter of a circle. Use the π button on your calculator to solve a problem involving π.
Supplementary Angles $m\angle 1 + m\angle 2 = 180°$	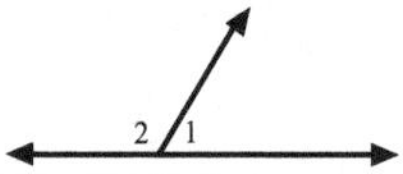	2 angles whose sum is 180°. (Need not be adjacent.)
Complementary Angles $m\angle 1 + m\angle 2 = 90°$	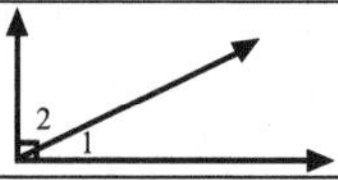	2 angles whose sum is 90°. (Need not be adjacent.)
Base $\overline{CD}$ is the base in parallelogram $DCBA$ and $\triangle CDE$. $\overline{CD}$ and $\overline{AB}$ are both bases in trapezoid $DCBA$ and h is the altitude.	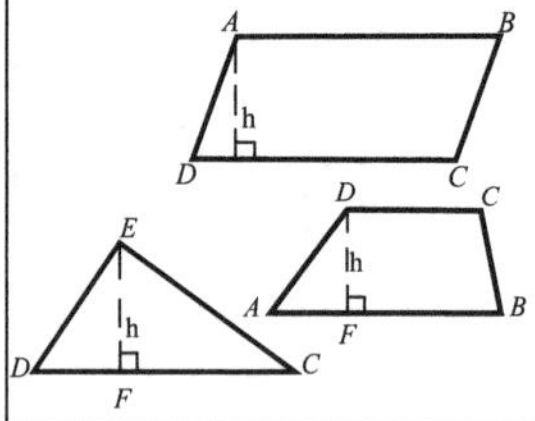	In a formula requiring a height measurement, the base of the polygon is the side that the altitude is drawn to. The formula for the area of a trapezoid uses both bases.
Altitude and Height $\overline{EF}$ is the altitude in the figure and measures the height.	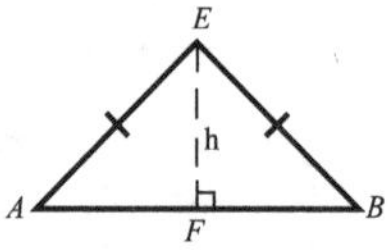	Altitude $\overline{EF}$ is the segment drawn from a vertex that is ⊥ to the opposite side.
Circle ⊙P	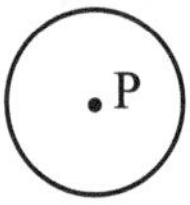	All the points in a plane at a given distance from a given point called the center. A circle is named by the center.
Arc $\overparen{CD}$	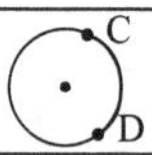	Part of a curve between two points.
Diameter	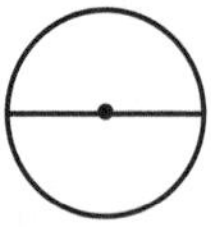	A line segment whose endpoints are on the circle and passes through the center.
Radius		Distance from the center of a circle to any point on the circle (the edge).
Interior Angle of a polygon $C, D, E,$ & F are all interior $\angle$'s.	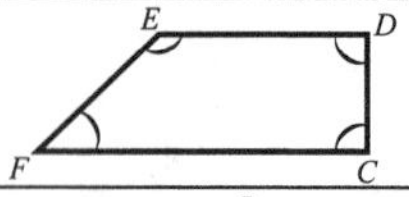	An angle formed on the inside of a polygon where the sides intersect.
Exterior Angle of a polygon $\angle BCD$ is exterior.	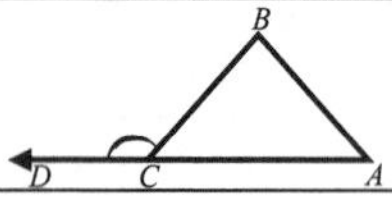	An outside angle formed by extending the side of the figure.
Vector		A directed line segment. It has magnitude represented by its length and direction shown by an arrow.

CONGRUENCE, PROOFS AND CONSTRUCTIONS

- Experiment with transformations in the plane.

- Understand congruence in terms of rigid motions.

- Prove geometric theorems.

- Make geometric constructions.

TRANSFORMATIONAL GEOMETRY TERMS

Transformational Geometry: A change in the position of points (or a figure) in a plane. It is a one-to-one mapping of the points in a pre-image to an image based on the transformation indicated. Some transformations are called "rigid motions" and some are called "non-rigid motions".

Mapping: A mapping in transformational geometry pairs each point in the pre-image with a point in its image based on the transformation indicated. Symbol: $\rightarrow$

Pre-image: The original figure that is to be transformed.

Image: The figure that is formed when each point in an original figure is moved in a specific way (mapped).

Vector: A directed line segment that indicates magnitude and direction. Vectors are used to show the distance and direction of a mapping from the pre-image to the image.

Composite Transformation: More than one transformation is performed on a figure. (See page 28)

Prime('): A "prime", is used to associate the image point with its pre-image point. The image of point A is A'. For multiple transformations, double ($''$) and triple primes ($'''$) are used to show subsequent images.

Rigid Motion: In rigid motion, distance and angle measure are both preserved. The pre-image and the image are congruent. *Translations, reflections, and rotations are rigid motions*. In addition to distance and angle measure being preserved, *betweenness* (a point between two others remains between those two points in the image) and *collinearity* (two or more points are located on a given line in the pre-image and the image) are preserved. A rigid motion transformation is also called an isometry.

Non-Rigid Motion: In a *dilation*, called a non-rigid motion, distance is not preserved, but angle measure is preserved. The pre-image and its image are similar. Corresponding angles are congruent and corresponding sides are proportional. *Betweenness* and *collinearity* are preserved.

Invariant: Points or properties that do not change when the pre-image is mapped onto its image. A point or points on the pre-image may be unchanged by the mapping over or under the arrow showing the mapping.

Notation or Rule: The type of transformation required is indicated by using a letter to show the type of transformation, and the additional information needed is given with it. There are examples shown in each description below. Sometimes the notation is done in front of the problem, sometimes it is over or under the arrow showing the mapping. The transformation itself can be written as a general rule using x and y and showing how they change as the transformation is performed.

Orientation (order): If the pre-image has a clockwise direction of the labels on its points and the image has the same clockwise direction of its labels, they have the same or direct orientation. (Rotations, dilations, translations have the same orientation.) Opposite *or* indirect *or* reverse orientation *or* reverse order means that the pre-image and its image have opposite directions of their labels. If the pre-image is clockwise, the image is counter-clockwise. (Line refections and glide reflections have opposite or reverse orientation.)

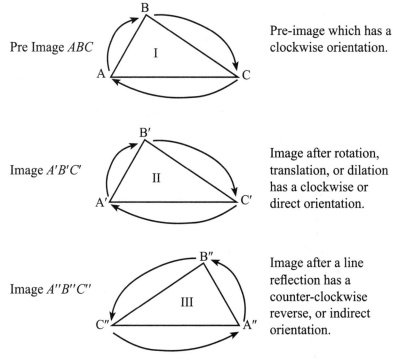

Pre Image *ABC*

Pre-image which has a clockwise orientation.

Image *A'B'C'*

Image after rotation, translation, or dilation has a clockwise or direct orientation.

Image *A''B''C''*

Image after a line reflection has a counter-clockwise reverse, or indirect orientation.

The pre-image I and image II have the same orientation or direct orientation. Both are clockwise.

The pre-image I and image III have reverse, opposite, or indirect orientation. One is clockwise, the other counter-clockwise.

RIGID MOTIONS

Rigid Motion Transformations include reflections, rotations, and translations. Distance and angle measure are preserved. The pre-image and its image are congruent.

REFLECTIONS

Line Reflection (*r*): Under a line reflection on line *m*, if a point, *A* is connected to its image, *A'*, a line segment *AA'* is formed and line *m* is the perpendicular bisector of *AA'*. In a line reflection, if the diagram is folded exactly on the line of reflection and the two halves are put together and held up to the light, the original figure and its image will be an exact match. Each side of the line of reflection is a "mirror image" of the other side. △*A'B'C'* is the image of △*ABC* under a line reflection on line *m*.

In symbols: $r_m(\triangle ABC) = \triangle A'B'C'$ *or* $\triangle ABC \xrightarrow{\ r_m\ } A'B'C'$

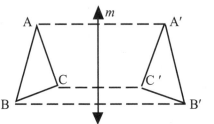

Properties under a line reflection: Distance, angle measure, and collinearity are invariant. Any of these properties that are true for the pre-image will remain unchanged in its image. Orientation in a line reflection is reversed or opposite. A line reflection is a rigid motion and is also called an opposite isometry.

Examples

❶ $r_{y\text{-}axis}$: If the line of reflection is the *y*-axis and the coordinates of the original point are (*x*, *y*), then the coordinates of its image will be (–*x*, *y*).

$$(x, y) \xrightarrow{\ r_{y\text{-}axis}\ } (-x, y)$$

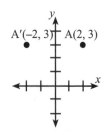

---

❷ $r_{x\text{-}axis}$: If the coordinates of the original point are (x, y), then the coordinates of its image will be $(x, -y)$.

$$(x, y) \xrightarrow{\;r_{x\text{-}axis}\;} (x, -y)$$

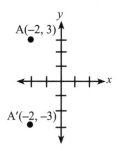

❸ $r_{y=x}$: If the line of reflection is the line representing the equation $y = x$ and the original point is (x, y), then its image is (y, x).

$$(x, y) \xrightarrow{\;r_{y=x}\;} (y, x)$$

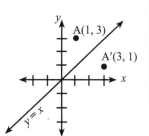

Congruence and Transformations

The line of reflection can be outside a figure or it may be part of the figure. It can be a line of symmetry when it is inside the figure, as shown in Figure 1.

Figure 1

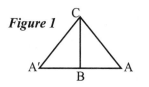

In this case, B and C are invariant points. They are their own images because they are on the line of reflection.

$$\triangle ABC \xrightarrow{\;r_{\overline{BC}}\;} \triangle A'BC$$

$$\overline{AC} \cong \overline{A'C} \qquad \angle CBA \cong \angle CBA'$$

$$\overline{AB} \cong \overline{A'B} \qquad \angle A \cong \angle A'$$

$$\overline{BC} \cong \overline{BC} \qquad \angle BCA \cong \angle BCA'$$

$$\triangle ABC \cong \triangle A'BC$$

Note: $\overline{BC}$ is unchanged by this reflection. $\overline{BC}$ is invariant.

Point Reflection (*r*): If a point A is reflected through point P, then point P will be the midpoint of the segment connecting point A and and its image, point A'. $\triangle ABC$ is reflected through point P. $\triangle A'B'C'$ is its image.

Symbols: $r_P(\triangle ABC) = \triangle A'B'C'$

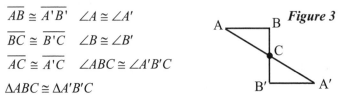

Figure 2

$\quad$ *or* $\triangle ABC \xrightarrow{r_P} \triangle A'B'C'$

$\overline{AC} \cong \overline{A'C'} \quad \angle A \cong \angle A'$

$\overline{AB} \cong \overline{A'B'} \quad \angle B \cong \angle B'$

$\overline{BC} \cong \overline{B'C'} \quad \angle C \cong \angle C'$

$\quad\quad \triangle ABC \cong \triangle A'B'C'$

Properties: Distance, angle measure, parallelism, collinearity are all preserved through a point reflection. Orientation is the same in both figures - counterclockwise in this example. It is a direct isometry.

A figure can be reflected through a point that is also part of the figure. In the accompanying diagram, it shows a reflection through point C.

$\overline{AB} \cong \overline{A'B'} \quad \angle A \cong \angle A'$

$\overline{BC} \cong \overline{B'C} \quad \angle B \cong \angle B'$

$\overline{AC} \cong \overline{A'C} \quad \angle ABC \cong \angle A'B'C$

$\triangle ABC \cong \triangle A'B'C$

Figure 3

Note: In this reflection, C is the point of reflection and is unchanged.

Summary: Reflections are rigid motion transformations. In both line and point reflections, the pre-image and the image are congruent. In a line reflection, the image has the opposite orientation (labeling is reversed) of the pre-image. A line reflection is an opposite isometry.

Note: **R** is sometimes used instead of **r** in reflections. Since the **R** for a rotation will include an angle reference, it is possible to distinguish between the two **R**'s. Follow your teacher's instructions for notation.

TRANSLATIONS

Translation (T): A shift of the figure in the x or horizontal direction and/or a shift in the y or vertical direction. Each point in the figure is translated the same distance and in the same direction. A vector is used to demonstrate the magnitude (distance) and direction of the translation. The value of the shift is added to the x and/or y coordinates of the point being translated to find the coordinates of its image. The rule can be written: $\triangle ABC \xrightarrow{T_{a,b}} \triangle A'B'C'.$ (See also page 148, Translation of a Circle.)

Example $\triangle ABC$ has vertices at $A(-5, 7)$, $B(-3, 4)$ and $C(-6, 1)$. After the translation, A' is located at $(4, 1)$. What are the coordinates of B' and C'? This requires finding the translation rule and applying it to B and C. Vectors can be used to show the magnitude (distance) and direction of the mapping of points A, B, and C, to A', B', and C' respectively. The magnitude of a vector can be found using the distance formula.

$A(-5, 7) \to A'(4, 1)$

The x changed from -5 to $+4$, so 9 was added to x.

The y changed from 7 to 1, so -6 was added to y.

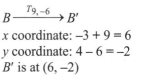

x coordinate: $-3 + 9 = 6$
y coordinate: $4 - 6 = -2$
B' is at $(6, -2)$

$C \xrightarrow{T_{9,-6}} C'$
x: $-6 + 9 = 3$
y: $1 - 6 = -5$
$C'(3, -5)$.

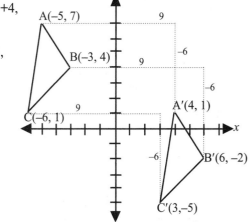

The rule could also be written: $T_{9,-6} \triangle ABC \to \triangle A'B'C'$, which means take the x and y values of A, B, and C and add 9 to each x value and add (-6) to each y value to find the coordinates of $A'B'C'$, the image of ABC. The size of the angles and the lengths of the sides in the image are congruent to the corresponding angles and sides in the pre-image.

$\angle A \cong \angle A'$	$\overline{AB} \cong \overline{A'B'}$	$\overline{AB} \parallel \overline{A'B'}$
$\angle B \cong \angle B'$	$\overline{BC} \cong \overline{B'C'}$	$\overline{BC} \parallel \overline{B'C'}$
$\angle C \cong \angle C'$	$\overline{AC} \cong \overline{A'C'}$	$\overline{AC} \parallel \overline{A'C'}$

Magnitude of the Vector: All 3 vectors have the same magnitude. Find the distance from A to A'. $d = \sqrt{(4-(-5))^2 + (1-7)^2} = \sqrt{81+36} = \sqrt{117}$

A translation is a rigid motion transformation because the pre-image and its image are congruent. In addition to distance, angle measure, collinearity, and betweenness being preserved, a translation also preserves parallelism. The orientation (labeling) is unchanged. A translation is a direct isometry.

Congruence and Transformations

ROTATIONS

Rotation (R): The turning of a figure around a point, called the center of rotation, for a specified number of degrees. Each point in the image is the same distance from the center of rotation as its corresponding point in the pre-image. The center of rotation can be part of the figure or it may be outside the figure, but it does not move in the rotation. If no point is mentioned, then the rotation is assumed to be about the origin. Rotations are <u>counterclockwise</u> unless otherwise stated. A negative degree marking indicates a clockwise rotation.

Figure 1

$R_{90°}$: 90° *counterclockwise* rotation about the origin

$\angle A \cong \angle A'$ $\overline{AB} \cong \overline{A'B'}$

$\angle B \cong \angle B'$ $\overline{BC} \cong \overline{B'C'}$

$\angle C \cong \angle C'$ $\overline{AC} \cong \overline{A'C'}$

$R_{B,-90°}$: −90° *clockwise* rotation around point B. This is equivalent to a 270° counterclockwise rotation. Point B does not move since it is the center of rotation.

Figure 2

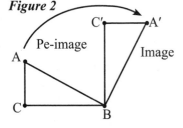

$\angle A \cong \angle A'$ $\overline{AB} \cong \overline{A'B}$

$\angle B \cong \angle B'$ $\overline{AC} \cong \overline{A'C'}$

$\angle C \cong \angle C'$ $\overline{BC} \cong \overline{BC'}$

Rules: When coordinates of a figure are given, the rules for rotations are as follows:

$$90°: (x, y) \xrightarrow{R_{90°}} (-y, x)$$

$$180°: (x, y) \xrightarrow{R_{180°}} (-x, -y)$$

$$270°: (x, y) \xrightarrow{R_{270°}} (y, -x)$$

$$360°: (x, y) \xrightarrow{R_{360°}} (x, y) \text{ Maps the figure onto itself.}$$

Note: Remember, rotations are counterclockwise unless the angle is negative. A rotation of −90° is the same as a rotation of 270°. −90° is a clockwise rotation, 270° is counterclockwise.

Properties: Distance, angle measure, betweenness, collinearity are all preserved under a rotation. It is a direct isometry.

2.2

Finding the Center of Rotation:

To find the center of rotation and the angle of rotation is a two step procedure. First construct the perpendicular bisectors of the segments joining two points in the pre-image with their corresponding points in the image. The intersection of the perpendicular bisectors is the center of rotation. A protractor can then be used to find the angle of rotation after connecting the point of rotation with one point and with its image.

Example Given $\triangle ABC$ and its image $A'B'C'$. Locate the center of rotation and measure the angle of rotation needed to map ABC to $A'B'C'$. Label the center P. Label the angle appropriately.

Congruence and Transformations

A) Find the center of rotation.
  1. Draw $\overline{BB'}$ and $\overline{CC'}$.
  2. Using a compass, draw the perpendicular bisectors of $\overline{BB'}$ and $\overline{CC'}$. Label them m and n respectively.
  3. The intersection for m and n is the center of rotation. Label it P.

Figure 1

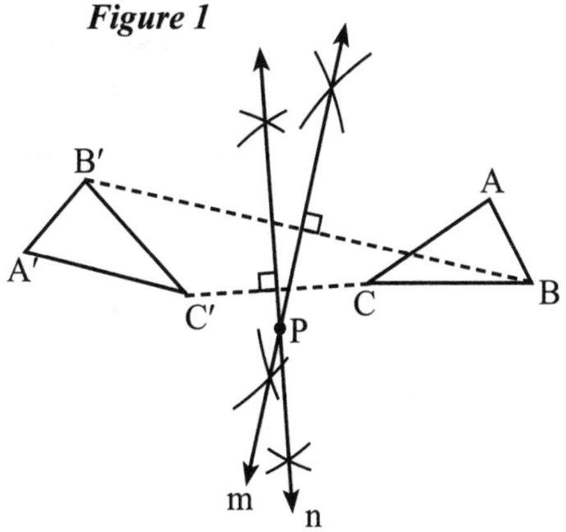

B) Find angle of rotation.
  1. Draw $\overline{PC}$ and $\overline{PC'}$ from P to form $\angle CPC'$.
  2. Measure $\angle CPC'$ with the protractor to find the angle of rotation.
  3. Label with the appropriate arc and degrees.

Figure 2

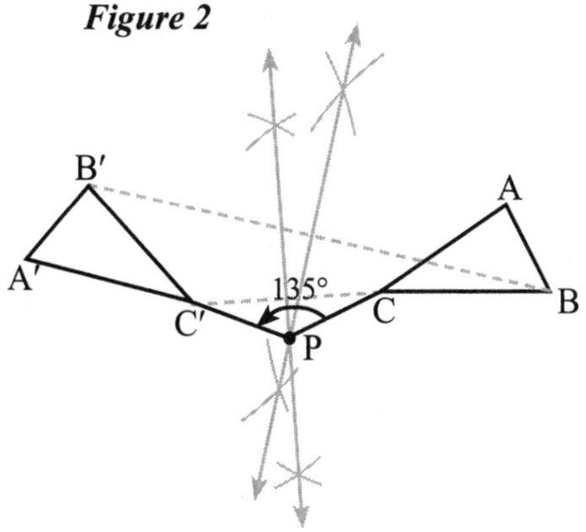

Conclusion: A rotation of 135° about point P is performed on $\triangle ABC$ to form $\triangle A'B'C'$.

In symbols this can be written: $\triangle ABC \xrightarrow{R_{P,135°}} \triangle A'B'C'$

SYMMETRY AND TRANSFORMATIONS
THAT MAP A FIGURE ONTO ITSELF

Mapped onto itself means the pre-image and the image coincide after being transformed.

Some figures have **line symmetry** or **rotational symmetry** and **sometimes point symmetry**. It has line symmetry if the figure that is reflected over a line maps the image onto the pre-image. **Rotational symmetry** occurs if a figure can be rotated around a point using a rotation greater than 0° and less than 360° which maps the figure onto itself. **Point symmetry** is a type of rotational symmetry and it occurs if a figure can be rotated 180° about a point to be mapped onto itself. Some figures do not have line or rotational symmetry.

Figure	Line of Symmetry and/or Points of Rotation	Degree of Rotational Symmetry	Line Symmetry
Isosceles Triangle		No rotational symmetry.	1 line
Equilateral Triangle		120°, 240°	3 lines
Scalene Triangle		No rotational symmetry.	No lines or points of symmetry.
Rectangle		180° This rotational symmetry is called point symmetry.	2 lines
Square		90°, 180°, 270°	4 lines

Congruence and Transformations

Figure	Line of Symmetry and/or Points of Rotation	Degree of Rotational Symmetry	Line Symmetry
Parallelogram		180° Has point symmetry.	No lines of symmetry.
Rhombus		180° Has point symmetry.	2 lines
Trapezoid		No rotational symmetry.	No lines or points of symmetry.
Isosceles Trapezoid		No rotational symmetry.	1 line
Regular Pentagon		72°, 144°, 216°, 288°	Lines of symmetry exist between each vertex drawn perpendicular to the opposite side. 5 lines

Congruence and Transformations

COMPOSITE TRANSFORMATIONS

Composite Transformations: More than one transformation can be performed on a figure. Multiple rigid motion transformations can be performed on a figure to create an image followed by that image being transformed again to form a 2nd image, and so on. Unlike some other procedures, in a composition of rigid motion transformations, the preservation of distance, angle measure, betweenness, and collinearity are maintained. In a composition of transformations it is necessary to perform the second transformation first, then perform the first transformation on that result.

Symbol ($\circ$) – Example: $r_{y\text{-}axis} \circ r_{x\text{-}axis}$: means reflect over the x-axis, then reflect that over the y-axis

Examples

❶ Given a triangle with vertices $A(2, 5)$, $B(3, 1)$, $C(7, 3)$. Find its image for $T_{-3, -5} \circ r_{y\text{–}axis}$. The pre-image and its images are congruent.

Steps

1) First perform the reflection over the y–axis of the triangle.

Rule: $r_{y\text{-}axis}(x, y) \rightarrow (-x, y)$

$A(2, 5) \xrightarrow{\ r_{y\text{-}axis}\ } A'(-2, 5)$

$B(3, 1) \xrightarrow{\ r_{y\text{-}axis}\ } B'(-3, 1)$

$C(7, 3) \xrightarrow{\ r_{y\text{-}axis}\ } C'(-7, 3)$

2) Then use those results to perform the translation.

Rule: $T_{-3, -5}(x, y) \rightarrow (x - 3, y - 5)$

$A'(-2, 5) \xrightarrow{\ T_{-3, -5}\ } A''(-5, 0)$

$B'(-3, 1) \xrightarrow{\ T_{-3, -5}\ } B''(-6, -4)$

$C'(-7, 3) \xrightarrow{\ T_{-3, -5}\ } C''(-10, -2)$

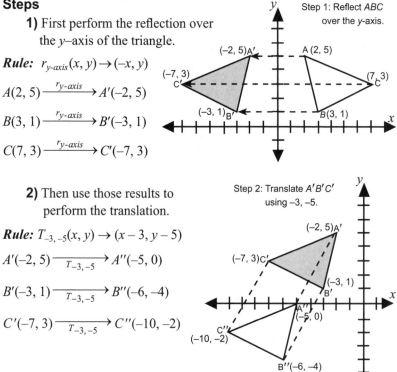

Conclusion: Both the reflection and the translation are rigid motion transformations. Distance and angle measure is preserved.

Therefore $\triangle ABC \cong \triangle A''B''C''$

---

❷ The diagram below shows a composite transformation of triangle *ABC* to triangle *A″B″C″*. Arrows are drawn to show the sequence of the steps.

a) What sequence of transformations is used?

b) Is △*ABC* congruent to △*A″B″C″*?

c) Justify your answer.

Solution:

a) This is a translation of △*ABC* where $(x, y) \rightarrow (x - 10, y - 5)$ is followed by a rotation of 90° in which $(x, y) \rightarrow (-y, x)$.

b) △*ABC* ≅ △*A″B″C″*

c) The pre-image is translated and then rotated. Both are rigid motions, therefore the pre-image and its images are congruent.

❸ In the diagram below, triangle *B* is the image of triangle *A*. Triangle *C* is the image of triangle *B* after being transformed.

a) Write a series of transformation to describe this procedure.

b) Write a single transformation that could be used to transform triangle *A* to its image triangle *C*.

Solution:

a) Triangle *A* is reflected over the *x*-axis to create triangle *B*. Triangle *B* is then reflected over the *y*-axis to create triangle *C*.

$$\triangle A \xrightarrow{r_{x\text{-}axis}} \triangle B$$

$$\triangle B \xrightarrow{r_{y\text{-}axis}} \triangle C$$

or

$$\triangle A \xrightarrow{r_{y\text{-}axis} \circ r_{x\text{-}axis}} \triangle C$$

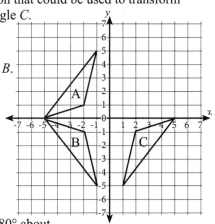

b) Triangle *A* can be rotated 180° about the origin to create triangle *C*.

$$\triangle A \xrightarrow{R_{0, 180°}} \triangle B$$

Glide Reflection: A composite transformation in which a figure is reflected through a line and is translated along that line in a direction parallel to the line of reflection. In this particular composite transformation, the order of the transformations is not important. However, in order to avoid confusion, it is best to continue to work in the same order as we do for other composite transformations. Start at the right and work backwards.

Example Triangle ABC has vertices $A(2,-1)$, $B(-3,-2)$ and $C(-1,-3)$. Sketch and give the coordinates of the image of a glide reflection performed on $\triangle ABC$ of $T_{-4,0}$ and $r_{x\text{-}axis}$.

Steps

1) Do the reflection.

Rule: $r_{x\text{-}axis}\,(x, y) \rightarrow (x, -y)$

$A(2, -1) \xrightarrow{\;r_{x\text{-}axis}\;} A'(2, 1)$

$B(-3, -2) \xrightarrow{\;r_{x\text{-}axis}\;} B'(-3, 2)$

$C(-1, -3) \xrightarrow{\;r_{x\text{-}axis}\;} C'(-1, 3)$

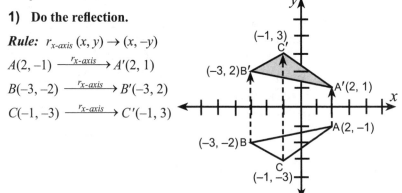

2) Do the translation.

Rule: $T_{-4,\,0} \rightarrow (x - 4, y)$

$A'(2, 1) \rightarrow A''(-2, 1)$

$B'(-3, 2) \rightarrow B''(-7, 2)$

$C'(-1, 3) \rightarrow C''(-5, 3)$

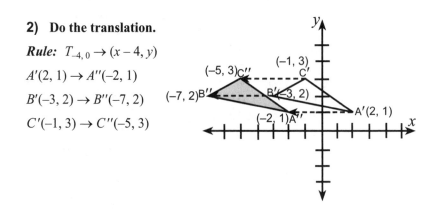

3) Conclusion: The coordinates of ABC after the glide reflection are $A''(-2, 1)$, $B''(-7, 2)$, and $C''(-5, 3)$ $\triangle ABC \xrightarrow{\;\;T_{-4,\,0}\,\circ\,r_{x\text{-}axis}\;\;} \triangle A''B''C''$

Note: Since the reflection and the translation are both isometries, the glide reflection is an isometry also.

TRANSFORMATION OF FUNCTIONS

Functions as well as geometric figures can be transformed. If the transformation is a rigid transformation, the pre-image and its image will be the same size and shape. Although it may not involve segments or angles, the pre-image and its image are congruent.

Example Determine the coordinates of the points in the following function, $f(x)$ after a reflection over the x-axis. Write the transformation in function form, describe the rule for the transformation, make a table of values for $g(x)$, and sketch the pre-image $f(x)$ and its image $g(x)$.

$$f(x) \xrightarrow{\quad r_{x\text{-axis}} \quad} g(x)$$

x	$f(x)$
–2	6
–1	3
0	2
1	3
2	6

In a reflection over the x-axis, (x, y) becomes $(x, -y)$.

x	$g(x)$
–2	–6
–1	–3
0	–2
1	–3
2	–6

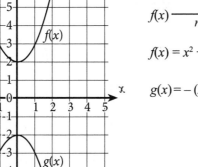

Written in the function form, this reflection would be:

$$f(x) \xrightarrow{\quad r_{x\text{-axis}} \quad} g(x)$$

$$f(x) = x^2 + 2$$

$$g(x) = -(x^2 + 2) \ or \ g(x) = -x^2 - 2$$

Congruence and Transformations

GEOMETRIC PROOFS

Proof: A method of demonstrating the logical sequence of acceptable mathematical reasoning steps used to reach a conclusion. In a problem, certain information is "given" and is accepted as being true. The given information is also called a premise or hypothesis. The solving of the problem, or the "proof", enables us to reach a conclusion - either stated in the problem or determined by analyzing the problem. In order to be certain that the conclusion reached is a correct one, we need to follow logical reasoning steps to get there. We can use hypotheses, axioms, definitions, and proven theorems to demonstrate our thinking process in reaching the conclusion. When doing proofs, it is helpful to note the information you are given on the diagram and mark each part that you prove as you go along. This strategy gives a better picture of the progression of the proof. At times, it is necessary to change the way you are doing a problem if you see that the steps you have completed are not leading to the conclusion desired. Re-planning the approach to the problem may be needed. Start over!

Note: It is important to realize that there are often several ways to develop the logical steps in a proof.

Types of Proofs: Many different forms of proofs can be used. They include Euclidean proofs (which are also called statement-reason or two column proofs), paragraph, flowchart, indirect, analytic or coordinate proofs, and formulas. A diagram demonstrating the problem often accompanies the proof. In each type of proofs, a statement is made and is supported by an associated reason that justifies it.

Diagram: A figure that represents the information given in the problem. Labels are very important. Although it doesn't need to be drawn to scale, it is best if the diagram is a fairly good representation of the figure – it will help you think of ideas about solving the problem. As information is developed in the problem, add that information to the diagram – angle or side congruency, perpendicularity, etc., the diagram should be shown as part of the proof on your paper – or as part of the solution to a problem. It can replace a "let" statement in some problems.

Statement: A step that is used in working toward a conclusion in a proof.

Reason: The supporting information for a statement.

REASONS AND TERMINOLOGY

The following are some of the mathematical statements that can be used as the supporting reason in a proof. Please note that *in this handbook*, due to the inconsistency of labeling between texts, instead of naming a theorem, axiom, corollary, etc., the reason is written but not labeled. Follow the directives of your teacher.

Axiom or Postulate: A proposition that is a starting point from which further logical steps can be determined. These are statements that are accepted without proof. Axioms or postulates are used to prove theorems. (Axiom originally referred to equal and unequal terms and postulate referred to geometric figures. Today they mean the same thing. Some texts use axiom and some use postulate.)

Ex: If two parallel lines are cut by a transversal, corresponding angles are congruent.

Theorem: A proposition that can be proved logically using principles of deduction.

Ex: The sum of the angles of a triangle is 180°.

Corollary: A theorem that follows from another theorem. A corollary can be easily deduced from another theorem. Some texts label these as theorems, some call them corollaries.

Ex: Each angle of an equiangular triangle has a measure of 60°.

Reason: Justification for using a statement. Reasons can be *Postulates - Axioms - Theorems - Corollaries - Properties - Definitions - Formulas*: These are all mathematically acceptable "reasons" which can be used to solve geometry problems and produce geometry proofs. Some texts name things differently - axiom vs. postulate, etc. In this handbook we will usually just call them "reasons". Each student should *classify and use the reasons according to the preferences of his/her teacher and using his/her own textbook.*

Proofs

Congruent vs. Equal: Congruent (≅) refers to the shape and size of a geometric figure. Congruent figures have the same shape and the same size. Their angles have the same measure and their sides have the same length. Equal (=) refers to the measure of a figure. These terms are used fairly interchangeably in many situations. Even the theorems, which are standard, are written in some texts using the word "congruent" and in others using the word "equal".

When discussing the congruence of line segments, a "hat" is used over the segment. $\overline{AC} \cong \overline{DE}$.
Congruence of angles is shown like this: $\angle ABC \cong \angle DEF$

The length of a segment is shown as: $AC = 10$.
The measure of an angle(s) uses an "m" with the angle symbol: $m\angle A = 75$

- If $AC = 10$ and $DE = 10$, then $\overline{AC} \cong \overline{DE}$.
- If $m\angle A = 75$ and $m\angle B = 75$, then $\angle A \cong \angle B$.

Corresponding Parts: Angles or sides of a polygon that are in matching positions with another polygon (used with similar or congruent polygons).

Properties of Equality: These properties are used as reasons in deductive reasoning geometry proofs. (They are also used in algebra when working with real numbers.)

Addition Property of Equality: If $a = b$, and $c = d$, then $a + c = b + d$
If equal quantities are added to equal quantities, the sums are equal.

Subtraction Property of Equality: If $a = b$, and $c = d$, then $a - c = b - d$.
If equal quantities are subtracted from equal quantities, the differences are equal.

Multiplication Property of Equality: If $a = b$, then $ac = bc$.
If equal quantities are multiplied by equal quantities, the products are equal.

Division Property of Equality: If $a = b$ and $c \neq 0$, then $a \div c = b \div c$.
If equal quantities are divided by equal quantities which are $\neq 0$, the quotients are equal.

Reflexive Property of Equality: Anything is equal to itself.
[numbers, line segments, angles]

Symmetric Property of Equality: If $a = b$, then $b = a$.

Transitive Property of Equality: If $a = b$ and $b = c$, then $a = c$.

Substitution Property of Equality: If $a = b$, then a can be replaced by b and b can be replaced by a. [If $a + b = c$ and $m + n = c$, then $a + b = m + n$ using substitution.]

Partition Postulate: The whole is equal to the sum of its parts. This is also called "**Segment Addition**" or "**Angle Addition**" or "**Betweenness**" depending on its use in a problem.

Betweenness: If point B is between points A and C, then $\overline{AB} + \overline{BC} = \overline{AC}$.

Very commonly used definitions:

Distance:
1) The length of a segment between two points.
2) The length of the perpendicular segment from a point to a line. Also called the perpendicular distance. This is the "default" – always use the perpendicular distance unless directed otherwise.
3) Equidistant means "equally distant." Example: any two or more points on a circle are equidistant from the center.

Midpoint Definition: If point M, on a line segment $\overline{AC}$, divides the segment into two congruent segments, $\overline{AM} \cong \overline{MC}$, then point M is the midpoint of $\overline{AC}$.

Definition of Bisector:
1) **Angle Bisector:** An angle bisector divides an angle into two congruent angles.
2) **Bisector of a segment:** The bisector of a line segment divides a line segment into two congruent line segments.

Points and Lines in a Plane:
- Through any two points there is exactly one straight line.
- If two lines intersect in a plane, their intersection is a point.
- If two points lie in a plane, then the line joining them lies in that plane.
- In a plane, two lines that do not intersect are parallel lines.
- In a plane, two lines that intersect to form right angles are perpendicular.
- Through any three noncollinear points there is exactly one plane.
- Through any point not on a given line, there exists one and only one line parallel to the given line in that plane.
- Through any point not on a given line, there exists one and only one line perpendicular to the given line in that plane.

PROVING GEOMETRIC THEOREMS

There are basic theorems that are used to develop the reasons in many types of geometry problems. Understanding the logical steps that justify the theorems is helpful. It is important to note that there are usually several ways to develop the logical steps in a geometry proof.

Procedure:

1) Carefully make and label a diagram based on the given information. The diagram is considered as "given."

2) Use the written "given" information as the first step. The reason is "Given."

3) Continue from there, basing the next statement on the ones before it which can include the diagram. The associated reason must be mathematically acceptable (definitions, theorems, axioms, corollaries, properties, etc) and it is written out in words or symbols.

4) Continue in a logical sequence, always making a statement based on prior steps, until the conclusion is reached. The conclusion is the "Prove" part of the problem.

5) The final step should state what has been proven (the conclusion) with its accompanying reason.

THEOREMS ABOUT LINES AND ANGLES

Intersecting Lines

Theorem: Vertical angles are congruent.

Given: $\overline{AB}$ intersects $\overline{CD}$ at E.

Prove: $\angle AEC \cong \angle BED$; $\angle AED \cong \angle BEC$

Statement-Reason Proof

Statement	Reason
1. $\overline{AB}$ intersects $\overline{CD}$ at E	1. Given.
2. $\angle 1 + \angle 2 = 180°$ $\angle 2 + \angle 3 = 180°$ $\angle 3 + \angle 4 = 180°$ $\angle 4 + \angle 1 = 180°$	2. If two angles share one common side and form a straight line, they are supplementary and supplementary angles total 180°.
3. $\angle 1 + \angle 2 = \angle 2 + \angle 3$ $\angle 2 + \angle 3 = \angle 3 + \angle 4$	3. Substitution.
4. $\angle 1 = \angle 3$; $\angle AEC \cong \angle BED$ $\angle 2 = \angle 4$; $\angle AED \cong \angle BEC$	4. Subtraction.

Geometry Made Easy – Common Core Standards Edition

Proofs

Parallel Lines Cut by a Transversal

Theorem: When a transversal crosses parallel lines, alternate interior angles are congruent.

To prove this theorem, we will use an axiom which states: When two parallel lines are cut by a transversal the corresponding angles are congruent. An axiom cannot be proved but is accepted as true.

Note: In this diagram, the angles are labeled with numbers to make it easier to follow. Angles 3 & 5, and 4 & 6 are the alternate interior angles. Corresponding angles are 1 & 5, 2 & 6, 3 & 7, 4 & 8.

Given: $M \parallel N$; L is a transversal.

Prove: $\angle 3 \cong \angle 5$

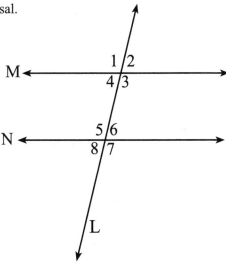

Paragraph Proof:
Given information includes lines M and N that are parallel and are cut by a transversal. Angle 1 and angle 5 are corresponding angles by the definition of corresponding angles. $\angle 1 \cong \angle 5$ because corresponding angles are congruent. $\angle 1$ and $\angle 3$ are vertical angles by the definition of vertical angles and therefore, $\angle 1 \cong \angle 3$. Since $\angle 5$ and $\angle 3$ are both congruent to a third angle, $\angle 1$, they are congruent to each other. Angles 5 and 3 are alternate interior angles and are congruent to each other ($\angle 3 \cong \angle 5$).

Theorem: Points on the perpendicular bisector of a line segment are equidistant from the segment's endpoints.

Given: C is on the perpendicular bisector of $\overline{AB}$.

Prove: $\overline{AC} \cong \overline{BC}$

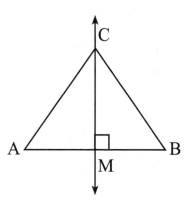

Paragraph Proof:
The given includes segment $\overline{AB}$ and point C on the perpendicular bisector of $\overline{AB}$. $\overline{AC}$ and $\overline{BC}$ are drawn. The intersection is labeled M. M is the midpoint of $\overline{AB}$ based on the definition of perpendicular bisector. $\overline{AM} \cong \overline{MB}$ since a midpoint divides a segment into two congruent parts. $\overline{CM} \cong \overline{CM}$ using the reflexive property. $\angle CMA$ and $\angle CMB$ are both right angles and are congruent because a perpendicular line creates two right angles and all right angles are congruent. $\triangle CMA \cong \triangle CMB$ by SAS $\cong$ SAS. Corresponding parts of congruent triangles are congruent (sometimes abbreviated CPCTC) making $\overline{AC} \cong \overline{BC}$.

- This proof can also be done using the properties of rigid motion transformations.

$\overline{CM}$ is the line of reflection which transforms $\overline{AM}$ to $\overline{MB}$. $\overline{AM} \cong \overline{MB}$ because a reflection preserves distance. C and M are mapped onto themselves because a point on the line of reflection remains unchanged. $\overline{AC} \cong \overline{BC}$ since distance is preserved in a reflection.

THEOREMS ABOUT TRIANGLES

Theorem: The measures of the interior angles of a triangle have a sum of 180°.

Given: $\triangle ABC$

Prove: $\angle 1 + \angle 2 + \angle 3 = 180°$

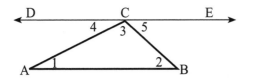

Statement Reason Proof:

Statement	Reason
1. $\triangle ABC$	1. Given.
2. Through C, draw a line, $\overline{DE}$, parallel to $\overline{AB}$. (Label the angles as shown on the diagram.)	2. Through a point not on a line, one and only one line parallel to the given line can be drawn. Angles are labeled for convenience.
3. $m\angle 4 + m\angle ECA = 180°$	3. Two adjacent angles whose sides form a straight line are supplementary and have a sum of 180°.
4. $m\angle ECA = m\angle 5 + m\angle 3$	4. Angle Addition.
5. $m\angle 4 + m\angle 3 + m\angle 5 = 180°$	5. Substitution.
6. $\angle 1 \cong \angle 4$; $\angle 2 \cong \angle 5$	6. When parallel lines are cut by a transversal, alternate interior angles are congruent.
7. $m\angle 1 + m\angle 2 + m\angle 3 = 180°$	7. Substitution.

Theorem: Base angles of an isosceles triangle are congruent.
(Or: If two sides of a triangle are congruent, the angels opposite them are congruent.)

Given: $\triangle ABC$ with $\overline{AB} \cong \overline{AC}$.

Prove: $\angle C \cong \angle B$

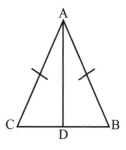

Statement Reason Proof

Statement	Reason
1. $\triangle ABC$ with $\overline{AB} \cong \overline{AC}$.	1. Given.
2. Construct $\overline{AD}$ so it bisects $\angle CAB$.	2. An angle has one and only one bisector.
3. $\angle CAD \cong \angle BAD$	3. Definition of angle bisector.
4. $\overline{AD} \cong \overline{AD}$	4. Reflexive property.
5. $\triangle CAD \cong \triangle BAD$	5. If 2 sides and the included angle of a triangle are congruent to the corresponding two sides and included angle of another, the triangles are congruent. (SAS $\cong$ SAS)
6. $\angle C \cong \angle B$	6. Corresponding part of congruent triangles are congruent. (CPCTC)

Theorem: The segment joining the midpoints of two sides of a triangle is parallel to the third side and equal to one-half its length.

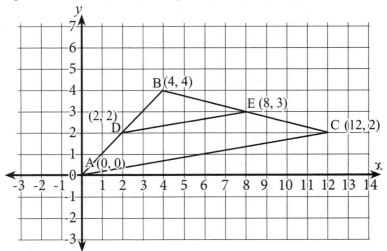

Given: $\triangle ABC$ with the mid-segment $\overline{DE}$. D is the midpoint of $\overline{AB}$, E is the midpoint of $\overline{BC}$. Coordinates of the vertices are labeled on the given diagram.

Prove: $\overline{DE} \parallel \overline{AC}$ $\overline{DE} = \frac{1}{2} \overline{AC}$

Coordinate Proof:

Plan: Determine that the slopes of $\overline{AC}$ and $\overline{DE}$ are equal proving they are parallel using the slope formula. Find the length of $\overline{AC}$ and $\overline{DE}$ and show that $\overline{DE} = \frac{1}{2} \overline{AC}$ using the distance formula.

Slope: $m = \dfrac{y_2 - y_1}{x_2 - x_1}$

$$m_{AC} = \frac{2-0}{12-0} = \frac{2}{12} = \frac{1}{6}$$

$$m_{DE} = \frac{3-2}{8-2} = \frac{1}{6}$$

Distance: $d = \sqrt{(x_2 - x_1)^2 + (y_2 - y_1)^2}$

$$d_{AC} = \sqrt{(12-0)^2 + (2-0)^2} = \sqrt{144+4} = \sqrt{148} = 2\sqrt{37}$$

$$d_{DE} = \sqrt{(8-2)^2 + (3-2)^2} = \sqrt{36+1} = \sqrt{37}$$

Conclusion: $\overline{AC} \parallel \overline{DE}$ because they have equal slopes. $\overline{DE} = \frac{1}{2} \overline{AC}$ as determined by the distance formula.

Theorem: The medians of a triangle meet at a point.

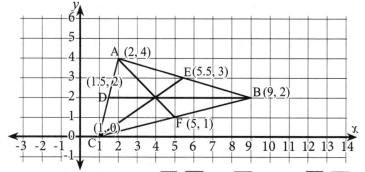

Given: $\triangle ABC$ with medians $\overline{CE}$, $\overline{AF}$, and $\overline{BD}$ drawn to $\overline{AB}$, $\overline{BC}$, and $\overline{AC}$ respectively. The coordinates of the points are labeled on the given diagram.

Prove: $\overline{CE}$, $\overline{AF}$ and $\overline{BD}$ intersect at one point.

Coordinate Proof:

Plan: Find the equations for two of the medians using the point slope formula. Set the two equations equal to each other to determine the coordinates of their intersection. Find the equation of the third median and test to see if the results found in the earlier step are on that line.

Formulas: $m(x_2 - x_1) = (y_2 - y_1)$		
$\overline{AF} : m(2 - 5) = (4 - 1)$	$\overline{CE} : m(5.5 - 1) = (3 - 0)$	$\overline{BD} : m(9 - 1.5) = (2 - 2)$
$-3m = 3$	$4.5m = 3$	$7.5m = 0$
$m = -1$	$m = \dfrac{3}{4.5} = \dfrac{2}{3}$	$m = 0$
$-1(x - 5) = (y - 1)$	$\dfrac{2}{3}(x - 1) = (y - 0)$	$0(x - 9) = (y - 2)$
$-x + 5 = y - 1$	$\dfrac{2}{3}x - \dfrac{2}{3} = y$	$y = 2$
$y = -x + 6$		

Solve the pair of equations for $\overline{AF}$ and $\overline{CE}$ simultaneously to determine where they intersect.

$\overline{AF} : y = -x + 6 \implies \overline{CE} : y = \dfrac{2}{3}x - \dfrac{2}{3} \implies -x + 6 = \dfrac{2}{3}x - \dfrac{2}{3}$

$\implies -3x + 18 = 2x - 2 \implies -5x = -20 \implies x = 4$

$\implies y = -x + 6 \implies y = -4 + 6 \implies y = 2$

The point of intersection is at (4, 2). Since the equation for $\overline{BD}$ is simply $y = 2$, the point (4, 2) is on $\overline{BD}$.

Conclusion: $\overline{CE}$, $\overline{AF}$ and $\overline{BD}$ intersect at one point, (4, 2).

THEOREMS ABOUT PARALLELOGRAMS

Theorem: Opposite sides of a parallelogram are congruent.

Given: Parallelogram $ABCD$ with diagonal $\overline{AC}$ drawn.

Prove: $\overline{AB} \cong \overline{CD}$, $\overline{BC} \cong \overline{AD}$

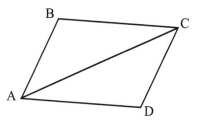

Flowchart Proof:

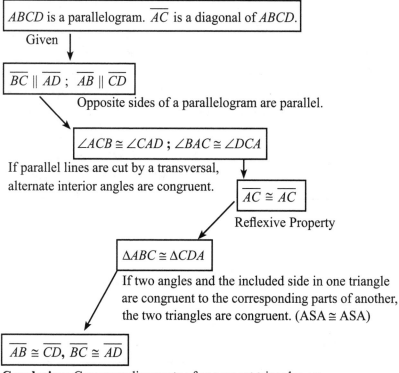

$ABCD$ is a parallelogram. $\overline{AC}$ is a diagonal of $ABCD$.

Given

$\overline{BC} \parallel \overline{AD}$; $\overline{AB} \parallel \overline{CD}$

Opposite sides of a parallelogram are parallel.

$\angle ACB \cong \angle CAD$; $\angle BAC \cong \angle DCA$

If parallel lines are cut by a transversal, alternate interior angles are congruent.

$\overline{AC} \cong \overline{AC}$

Reflexive Property

$\triangle ABC \cong \triangle CDA$

If two angles and the included side in one triangle are congruent to the corresponding parts of another, the two triangles are congruent. (ASA $\cong$ ASA)

$\overline{AB} \cong \overline{CD}$, $\overline{BC} \cong \overline{AD}$

Conclusion: Corresponding parts of congruent triangles are congruent. (CPCTC)

Proofs

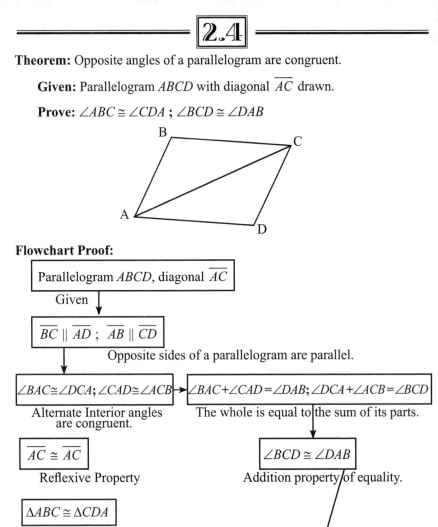

2.4

Theorem: Opposite angles of a parallelogram are congruent.

Given: Parallelogram $ABCD$ with diagonal $\overline{AC}$ drawn.

Prove: $\angle ABC \cong \angle CDA$; $\angle BCD \cong \angle DAB$

Flowchart Proof:

Parallelogram $ABCD$, diagonal $\overline{AC}$

Given

$\overline{BC} \parallel \overline{AD}$; $\overline{AB} \parallel \overline{CD}$

Opposite sides of a parallelogram are parallel.

$\angle BAC \cong \angle DCA$; $\angle CAD \cong \angle ACB$
Alternate Interior angles are congruent.

$\angle BAC + \angle CAD = \angle DAB$; $\angle DCA + \angle ACB = \angle BCD$
The whole is equal to the sum of its parts.

$\overline{AC} \cong \overline{AC}$
Reflexive Property

$\angle BCD \cong \angle DAB$
Addition property of equality.

$\triangle ABC \cong \triangle CDA$

If 2 angles and the included side of a triangle are congruent to the corresponding parts of another, the triangles are congruent. (ASA $\cong$ ASA)

$\angle ABC \cong \angle CDA$
Corresponding parts of congruent triangles are congruent.

$\angle ABC \cong \angle CDA$, $\angle BCD \cong \angle DAB$

Conclusion: Opposite angles of a parallelogram are congruent.

Proofs

2.4

Theorem: The diagonals of a parallelogram bisect each other.

Given: Parallelogram *ABCD*, diagonals
$\overline{AC}$ and $\overline{BD}$ intersect at *E*.

Prove: $\overline{AE} \cong \overline{EC}$, $\overline{BE} \cong \overline{ED}$

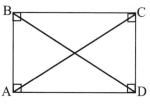

Paragraph Proof:
A parallelogram, *ABCD* is given along with diagonals $\overline{AC}$ and $\overline{BD}$ which intersect at *E*. $\overline{AB} \cong \overline{CD}$, $\overline{BC} \cong \overline{AD}$, $\overline{AB} \parallel \overline{CD}$, $\overline{BC} \parallel \overline{AD}$ because opposite sides of a parallelogram are congruent and parallel. $\angle DCE \cong \angle BAE$ and $\angle EBA \cong \angle EDC$ because when parallel lines are cut by transversal, alternate interior angles are congruent. That makes $\triangle ABE \cong \triangle CDE$ using the Angle-Side-Angle triangle congruency theorem. Corresponding parts of congruent triangles are congruent which makes $\overline{AE} \cong \overline{EC}$, $\overline{BE} \cong \overline{ED}$. The diagonals bisect each other because a bisector cuts a segment into two equal parts.

Proofs

Theorem: Rectangles are parallelograms with congruent diagonals.

Given: Rectangle *ABCD*

Prove: *ABCD* is a parallelogram,
$\overline{AC} \cong \overline{BD}$

Statement	Reason
1. *ABCD* is a rectangle.	1. Given.
2. $\angle A$, $\angle B$, $\angle C$ and $\angle D$ are right angles.	2. A rectangle contains 4 right angles.
3. $\angle A \cong \angle C$, $\angle B \cong \angle D$.	3. All right angles are congruent.
4. *ABCD* is a parallelogram.	4. A quadrilateral is a parallelogram if both pair of opposite angles are congruent.
5. $\overline{AB} \cong \overline{CD}$, $\overline{BC} \cong \overline{AD}$	5. Opposite sides of a parallelogram are congruent.
6. $\triangle ABC \cong \triangle CDA$	6. SAS triangle congruence.
7. $\overline{AC} \cong \overline{BD}$	7. Corresponding parts of congruent triangles are congruent. (CPCTC)

2.5

CONSTRUCTING LINES AND ANGLES

Construction: A drawing done in geometry using only a compass and a straight edge. Protractors, rulers, and graph paper or other devices for measuring are not permitted. No parts of a construction may be "sketched". Each part of a construction must be drawn with the compass and/or a straight edge. The lines and arcs used to make the construction are to be left on the paper. DO NOT ERASE the construction lines. In constructions, we are often duplicating the measure of something into a new drawing or we are using the compass to measure equal distances. If the compass should be changed from a previous step, it will be indicated. Also, the demonstration shown may not be the only way to do a construction. As we have seen throughout this geometry book, there are often several ways to reach the same result. Provided that acceptable construction methods are used with valid logical reasoning, other "steps" may be used to do a construction.

Justify or explain the construction: This means to discuss why a particular method works. Follow your teacher's instructions about specifics that may be required.

Examples Constructions

❶ Construct a line segment congruent to a given line segment.

> **Given:** $\overline{AB}$
>
> **Construct:** $\overline{CD} \cong \overline{AB}$

Steps:

1) Draw a line, m, in a different location than $\overline{AB}$.

2) Put the compass point on A and the pencil point on B. This measures the length of $\overline{AB}$.

3) Move the compass point to any point on line m and label the point C.

4) Swing an arc through line m.

5) Label the point of intersection of m and the arc, D.

6) $\overline{AB} \cong \overline{CD}$

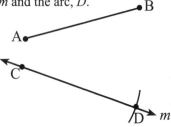

Discussion: By using the compass to measure the length of the given segment, we can mark off an equal segment on another line.

Geometry Made Easy – Common Core Standards Edition

❷ Construct an angle congruent to a given angle.

Given: ∠*ABC*

Construct: ∠*DEF* ≅ ∠*ABC*

Steps:

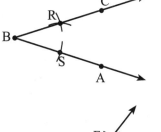

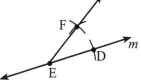

1) Draw a line *m* and place a point, *E* on the line.

2) Put the compass point on *B* in the given angle. Swing an arc through each side of ∠*ABC*. Label the points where the arcs intersect the sides of the angle with *R* and *S*.

3) Move the compass point to point *E* on line *m*.

4) Swing an arc that intersects *m* and label the point of intersection *D*.

5) With the point still at *E*, swing another arc in approximately the place that the other side of the angle will be.

6) Put the compass point on *R* and the pencil point on *S* to measure that distance.

7) Move the point of the compass to point *D*.

8) Swing an arc from point *D* to intersect with the arc from step 5. Label that point *F*.

9) Connect point *F* and point *E*.

10) ∠*DEF* ≅ ∠*ABC*

Discussion: If two angles are congruent, the measure between their sides at a specific distance from the vertex is equal. By marking off two equal segments on the sides of the given angle and measuring between the endpoints of the segments, we can determine the measures of corresponding parts of the congruent angle.

Constructions

❸ Construct the perpendicular bisector of a given line segment:

   Given: $\overline{AB}$
   Construct: $\overline{CD} \perp \overline{AB}$ and bisecting $\overline{AB}$

Steps::

1) Open the compass to more than 1/2 of $\overline{AB}$.

2) Put the compass point on A and swing two arcs - one above and one below $\overline{AB}$.

3) Move the compass point to B. Swing another pair of arcs, one above and one below $\overline{AB}$, so they intersect with the first pair.

4) Label the points where the arcs intersect, C and D.

5) Draw a line through point C and point D.

6) $\overline{CD}$ is the $\perp$ bisector of $\overline{AB}$.

Discussion: All the points on the perpendicular bisector of a line segment are equidistant from the endpoints of the segment. Swinging 2 pair of equal intersecting arcs, one pair below $\overline{AB}$ and the other above $\overline{AB}$, intersecting at C and D, creates 2 points that are each equidistant from points A and B. The perpendicular bisector of $\overline{AB}$ goes through C and D.

❹ Construct the bisector of a given angle.

Given: ∠ABC
Construct: $\overrightarrow{BD}$ so that $\overrightarrow{BD}$ bisects ∠ABC

Steps:

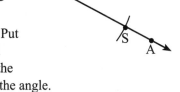

1) Put the compass point on *B* in the given angle. Swing an arc through each side of ∠ABC.

2) Label the points where the arcs intersect the sides of the angle with *R* and *S*.

3) Widen the compass a little. Put the compass point on *R* and swing an arc - put it out in the area toward the opening of the angle.

4) Move the compass point to point *S* and repeat step 3.

5) Label the point of intersection of the two arcs *D*.

6) Draw a line from point *B* through point *D*.

7) $\overrightarrow{BD}$ is the bisector of ∠ABC.

Discussion: All points on an angle bisector are equidistant from the sides of the angle. Locate two points *R* and *S*, one on each side of the angle, that are equally distant from the vertex. Swing a pair of equal intersecting arcs from each of those two points. The point of intersection, *D*, is equally distant from the sides the angle. Connect the intersection point with the vertex of the angle to form the angle bisector.

Constructions

❺ Construct a line perpendicular to a given line at a given point on the line.

Given: Line *m* and point *A* on line *m*.
Construct: $\overleftrightarrow{AB} \perp m$ at *A*.

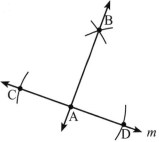

Steps:

1) Put the compass point on *A*.
Swing an arc on each side of
A through line *m*. Label the
points of intersection *C* and *D*.

2) Open the compass a little. Move
the point to *C* and swing an arc
above line *m*. Move the point to *D* an swing another arc above
line *m* so the two arcs intersect. Label the point of intersection
of the two arcs, *B*.

3) Connect *B* and *A*.

4) $\overleftrightarrow{AB} \perp$ line *m* at *A*. (Here we are making a segment on line
m and then constructing the perpendicular bisector of it to get
the perpendicular line required in the problem.)

Discussion: All the points on the perpendicular bisector of a line
segment are equidistant from the endpoints of the segment. Make
endpoints, *C* and *D*, by swinging equal arcs from point *A* that
intersect line *m*. Make another point, *B*, at the intersection of two
equal arcs from points *C* and *D*. *A* and *B* are each equidistant from
C and *D* so when *A* and *B* are connected, a line perpendicular to
line *m* at *A* is formed.

❻ Construct a line perpendicular to a given line from a given point not on the line.

> **Given:** Line *m* and point *C* not on line *m*.
> **Construct:** $\overleftrightarrow{CD} \perp m$.

Steps:

1) Put the compass point on *C*. Open the compass so the pencil point is on the opposite side of line *m*.

2) Swing an arc below line *m* so it intersects line *m* in two places, *E* and *F*.

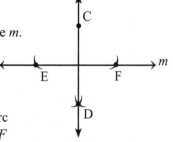

3) Put the point on *E* and swing an arc below *m*. Then move the point to *F* and swing another arc below *m* so it intersects with the previous arc. Label the point of intersection of the two arcs, *D*.

4) Connect point *C* and point *D*.

5) $\overleftrightarrow{CD} \perp m$. ($\overleftrightarrow{CD}$ is actually the perpendicular bisector of the segment $\overline{EF}$ which we created and used on line *m*.)

Discussion: All the points on the perpendicular bisector of a line segment are equidistant from the endpoints of the segment. Creating two points on *m* that are equidistant from *C* make two endpoints of a segment on line *m*. *D* is formed by swinging a pair of equal arcs from *E* and *F*. *D* is the second point that is equidistant from *E* and *F* and line $\overline{CD}$ is $\perp$ to line *m* through *C*.

Constructions

Constructions 7-8 are based on knowledge of the first 6 constructions.

❼ Construct a line parallel to a given line through a given point not on the line.

Given: Line $\overleftrightarrow{MN}$ and Point A not on $\overleftrightarrow{MN}$.
Construct: Line $S \parallel \overleftrightarrow{MN}$.

Plan: Use the knowledge that if two lines are cut by a transversal making a pair of corresponding angles that are congruent, the lines are parallel.

Steps:

1) Draw a line through point A to intersect with $\overleftrightarrow{MN}$ at point F creating $\angle MFA$. ($\angle 1$)(This line, $\overleftrightarrow{AF}$, will become the transversal.)

2) Using point A as the vertex of the angle and line $\overleftrightarrow{AF}$ as one side of the angle, construct $\angle 2$, congruent to $\angle 1$ with its vertex at A to make a pair of congruent corresponding angles. $\angle 1 \cong \angle 2$ (See construction 6)

3) The new side of the congruent angle is the line parallel to $\overleftrightarrow{MN}$. Extend that side and label it S.

4) $S \parallel \overleftrightarrow{MN}$

Discussion: Two parallel lines have corresponding angles that are congruent. By drawing a transversal through the line and also through the given point, the vertex and one side of the corresponding angle are formed. The second side of a congruent angle on the transversal at the given point results in a parallel line being created through the vertex.

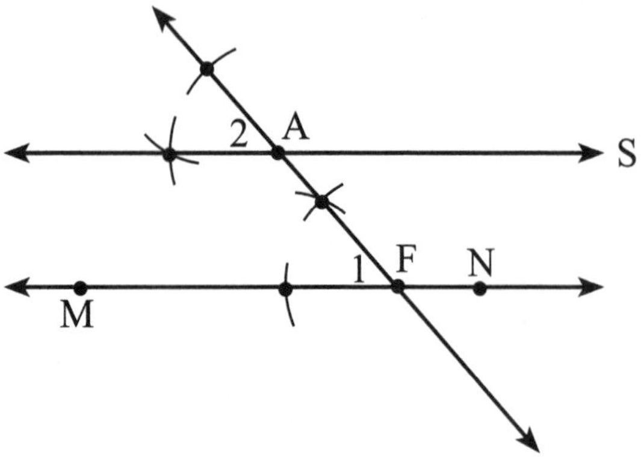

❽ Construct an equilateral triangle.

> **Given:** A segment congruent to each of the
> three congruent sides of the triangle.

Steps:

1) Draw a line and construct a line segment on it equal to the given segment.

2) Keeping the compass unchanged, create an arc from each endpoint of the new segment so the arcs intersect.

3) The point of intersection is the vertex of the equilateral triangle. Connect it to each endpoint of the segment.

Discussion: In an equilateral triangle, all three sides are congruent. Copy the length of the given segment with the compass and mark off a congruent segment for one side of the triangle. Swinging two more equal arcs gives a point of intersection that is the same distance from each endpoint. Connect the point of intersection to each endpoint to form two more congruent sides.

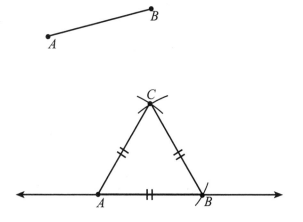

INSCRIBING POLYGONS IN A CIRCLE

A polygon inscribed in a circle is constructed so each vertex of the polygon is on the circle.

Examples Constructions

❶ Construct a square inscribed in a circle.

Given: Circle *O*.

Steps:

1) Draw diameter $\overline{AB}$.

2) Construct the perpendicular bisector of $\overline{AB}$ to make a 2nd diameter. Label it $\overline{CD}$.

3) Using the straightedge, connect the ends of the diameters to form quadrilateral *ADBC*.

Discussion: By constructing two perpendicular diameters, four right central angles are formed. Connecting the endpoints of the diameters creates 4 right triangles. Since the radii are all congruent, each right triangle is isosceles with base angles equal to 45°. Adding two base angles together makes each of the 4 angles at *A*, *D*, *B*, and *C* equal to 90° and, therefore right angles. The sides of the square are formed by the hypotenuse of each triangle. Since the triangles are congruent, the hypotenuses are congruent to each other because corresponding parts of congruent triangle are congruent. *ADBC* is a square.

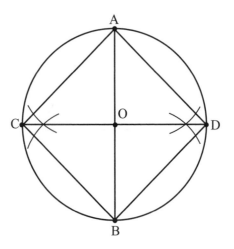

---

❷ Construct a regular hexagon in a circle.

Given: Circle O

Steps:

1) Draw a point, A, on the circle and measure the radius $\overline{OA}$. with a compass.

2) Starting at A, mark off along the circle arcs equal to the length of the radius. There will be 6 of them. Label the intersections with B, C, D, E, and F.

3) Connect A with B, B with C, and continue around the circle.

Figure 1 shows just the construction.

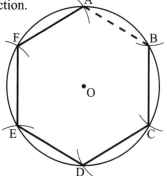

Figure 2 is sketched to show the analysis.

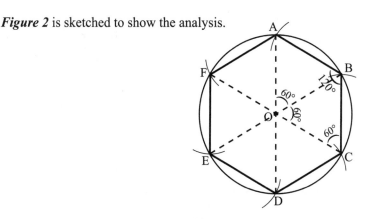

Discussion: Sketch radii $\overline{OA}$, $\overline{OB}$, and $\overline{OC}$. They are congruent to each other since they are radii of the same circle. $\overline{OA} \cong \overline{AB}$ since $\overline{OA}$ was used to measure $\overline{AB}$. $\triangle AOB$ is an equilateral triangle with 3 congruent sides. Equilateral triangles are also equiangular. Since two 60° angles are added together to form each angle of the hexagon, the angles of the hexagon are each 120°. Therefore it is a regular hexagon.

❸ Constructing an inscribed equilateral triangle in a circle.

Follow the same procedure as used in inscribing a regular hexagon in a circle. However, connect alternating points of intersection instead of every one. Connect *A* to *C*, *C* to *E*, and *E* to *A*. This makes a triangle, △*ACE*.

Figure 1 shows just the construction.

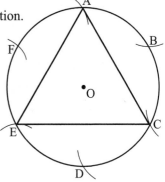

Figure 2 shows the analysis of the construction.

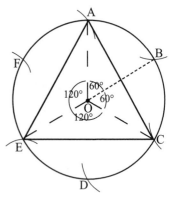

Discussion: Based on the construction of the hexagon on page 55, we can see that m∠*AOC* = 120°. $\overline{OA} \cong \overline{OC}$ because they are both radii and all radii of circle *O* are congruent. The same analysis can be done with each of the other two triangles. △*AOC* ≅ △*COE* ≅ △*EOA* based on the side-angle-side theorem of congruence. Therefore, $\overline{AC} \cong \overline{CE} \cong \overline{EA}$ because they are corresponding parts of congruent triangles. △*ACE* is an equilateral triangle inscribed in circle *O*.

Unit 3

SIMILARITY, PROOFS, AND TRIGONOMETRY

- Understand similarity in terms of similarity transformations.

- Prove theorems involving similarity.

- Define trigonometric ratios and solve problems involving right triangles.

- Apply geometric concepts in modeling situations.

Rigid motion transformations like rotations, reflections, and translations can be used to prove congruency and non-rigid motions like dilations can be used to prove similarity. (See also Unit 1)

Dilation: A dilation is a type of transformation that is a *non-rigid motion* and it does *not* preserve distance. Angle measure is preserved, but distance is changed based on the constant of dilation. Dilations can be used when showing that two figures are similar to each other.

A dilation is also called a similarity. The *x* and *y* coordinates of each point in the pre-image are multiplied by a constant to find the coordinates of its image. The center of dilation remains unchanged. The center of dilation is indicated in the notation if it is not the origin. A dilation of *k* where *k* is the constant of dilation, or scale factor, is a transformation such that:

    1. The image of point *O*, the center of dilation, is point *O*.

    2. The *x* and *y* coordinates of all other points are multiplied by *k* to find the coordinates of the images of the points.

Rule: Under a dilation of *k* whose center of dilation is the origin,
$$P(x, y) \xrightarrow{D_k} P'(kx, ky) \quad \underline{or} \quad D_k(x, y) = (kx, ky).$$

Example In the accompanying diagram, we have a dilation whose center of dilation is the origin and the constant of dilation is 2. For each point, multiply its *x* and *y* coordinates by 2 to get the coordinates of its image. Notice that the origin stays the same.

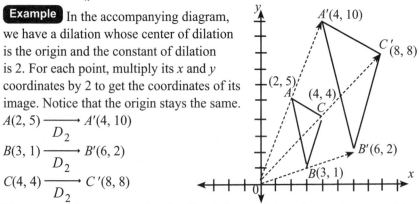

$$A(2, 5) \xrightarrow{D_2} A'(4, 10)$$

$$B(3, 1) \xrightarrow{D_2} B'(6, 2)$$

$$C(4, 4) \xrightarrow{D_2} C'(8, 8)$$

The image of a segment that is dilated is parallel to the pre-image and its length is in a ratio to the pre-image equal to the constant of dilation.

$$\overline{AB} \parallel \overline{A'B'} \; ; \; A'B' = 2\,AB$$

$$\overline{AC} \parallel \overline{A'C'} \; ; \; A'C' = 2\,AC$$

$$\overline{BC} \parallel \overline{B'C'} \; ; \; B'C' = 2\,BC$$

The angles *A'*, *B'*, and *C'* are congruent to their corresponding angles in the pre-image.

Note: If $0 < k < 1$ the image will be smaller than the pre-image.

If $k < 0$ the quadrant of the image will be different than the pre-image.

Properties: DISTANCE is *not* preserved under a dilation. The image and its original or pre-image are similar figures, but are not congruent (unless the constant of dilation is 1 in which case the figure is its own image). The other properties are still preserved: angle measure, collinearity, and betweenness. A dilation is *not* an isometry, it is a similarity.

Examples

❶ Determine the effect of a dilation on a line that passes through the center of dilation.

Given line *m* as shown. Perform a dilation of line *m* using the center of dilation at the origin and a constant of dilation, or scale factor, of 2. What conclusion can you draw?

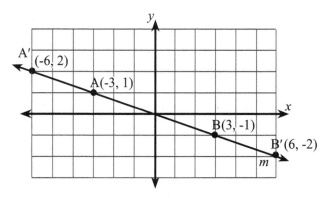

Solution: Pick two points *A* and *B* on *m* and determine their coordinates. Perform the dilation to find two image points. Sketch the line through the image points. Compare the pre-image and image.

When the sketch is drawn it is clear that *A'* and *B'* are on *m*. Since one and only one line can pass through two given points, *m* and its image are the same line.

Conclusion: When a line passing through the center of dilation is dilated, the line is unchanged. The image and the pre-image coincide.

Transformation and Similarity

❷ Given line *s* which contains the points *A* and *B*, and its image, line *t* which contains points *A'* and *B'*. What is the scale factor (constant of dilation) in this example? Where is the center of dilation? Discuss this in relation to the two lines and their relationship to each other.

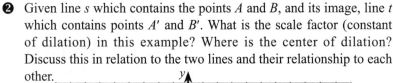

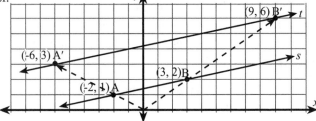

Solution: Since the coordinates of *A* are multiplied by 3 to obtain the coordinates of *A'* and the *B* coordinates are multiplied by 3 to obtain *B'*, the constant of dilation (or scale factor) is 3. The center of dilation is the origin. The lines appear to be parallel. This can be proven using the slope formula to see if the slopes of the two lines are equal. (The dotted line represents the vector showing the magnitude and direction of the dilation.)

Slope : $m = \dfrac{y_2 - y_1}{x_2 - x_1}$

$$m_s = \frac{2-1}{3-(-2)} = \frac{1}{5}$$

$$m_t = \frac{6-3}{9-(-6)} = \frac{3}{15} = \frac{1}{5}$$

$$\therefore s \parallel t$$

Conclusion: A line that does not pass through the center of dilation is parallel to its image after a dilation. Since the slopes of the lines are equal, the image and its pre-image are parallel. In the example above, the dilation can be written: $s \xrightarrow{\;D_3\;} t$ or $s \xrightarrow[D_3]{} t$

❸ Given Triangle *ABC* as shown. Using a scale factor of 0.5, sketch the image of *ABC* and label it *A'B'C'*. The center of dilation is at the origin. Is *A'B'C'* similar to *ABC*? Is there a relationship between the constant of dilation (scale factor), and the ratio of the sides? Justify your answer.

The coordinates of ABC are $A(3, 10)$, $B(8, 12)$, and $C(10, 6)$. Multiply the scale factor (0.5) by the coordinates of ABC to determine the coordinates of $A'B'C'$. They are $A'(1.5, 5)$, $B'(4, 6)$, and $C'(5, 3)$. Sketch $A'B'C'$ on the graph. To determine if the two triangles are similar, use the statement that if three sides of a triangle are proportional to the corresponding three sides of the other, the triangles are similar (SSS).

Use the distance formula to determine if the sides are proportional.

$\triangle ABC$

$$d_{AB} = \sqrt{(8-3)^2 + (12-10)^2} = \sqrt{25+4} = \sqrt{29}$$

$$d_{BC} = \sqrt{(10-8)^2 + (6-12)^2} = \sqrt{4+36} = \sqrt{40} = 2\sqrt{10}$$

$$d_{AC} = \sqrt{(10-3)^2 + (6-10)^2} = \sqrt{49+16} = \sqrt{65}$$

$\triangle A'B'C'$

$$d_{A'B'} = \sqrt{(4-1.5)^2 + (6-5)^2} = \sqrt{6.25+1} = \sqrt{7.25} = \sqrt{\frac{29}{4}} = \frac{\sqrt{29}}{2}$$

$$d_{B'C'} = \sqrt{(5-4)^2 + (3-6)^2} = \sqrt{1+9} = \sqrt{10}$$

$$d_{A'C'} = \sqrt{(5-1.5)^2 + (3-5)^2} = \sqrt{12.25+4} = \sqrt{16.25} = \sqrt{\frac{65}{4}} = \frac{\sqrt{65}}{2}$$

Are they proportional? Compare the ratios of the image to the pre-image to see if they are all equal.

$$\frac{A'B'}{AB} = \frac{.5\sqrt{29}}{\sqrt{29}} = .5 \text{ or } \frac{1}{2}$$

$$\frac{B'C'}{BC} = \frac{\sqrt{10}}{2\sqrt{10}} = .5 \text{ or } \frac{1}{2}$$

$$\frac{A'C'}{AC} = \frac{.5\sqrt{65}}{\sqrt{65}} = .5 \text{ or } \frac{1}{2}$$

Conclusion: Because the sides of one triangle are each proportional to the corresponding sides of the other, the triangles are similar. The ratio of the corresponding sides is one-half, and the constant of dilation (scale factor) is one-half. The proportionality of the image to the pre-image is also one-half.

Transformation and Similarity

❹ Use Example 3 to discuss the angles in *ABC* compared to the angles in *A′B′C′*. What can be determined about the triangles using this comparison? Explain your reasoning.

Solution: One of the properties of a dilation is that angle measure is preserved. Therefore, angles *A′*, *B′*, and *C′* are congruent to angles *A*, *B*, and *C* respectively. We can state that $\triangle ABC \sim \triangle A'B'C'$ since two triangles are similar if the three angles of one are congruent to the corresponding angles of the other.

❺ Explain why if a dilation is performed on a triangle and two angles of the pre-image triangle are congruent to the corresponding two angles of its image, the triangles must be similar.

Solution: Every triangle has 180° contained within its three interior angles. If two angles of a triangle are equal to two angles of another triangle and the sums of those angles are subtracted from 180°, the differences **are equal.** The third angle of each triangle must be equal also. This meets the similarity definition of AAA ≅ AAA (If 3 angles of one triangle are congruent to 3 angles of another, the triangles are similar).

COMPOSITE TRANSFORMATIONS

When two or more transformations are performed on a figure, the properties of each transformation are applied to the images created. Rigid motion transformations -- reflections, rotations, and translations -- can be combined and the final image will be congruent to the original figure. Remember that if figures are congruent, they are also similar. If any of the rigid motion transformations are combined with a dilation (a non-rigid motion transformation), the final image will be similar to the original. (***Note***: If the dilation has a scale factor of 1, the figures will still be similar but will also be congruent.) When performing composite transformations, work from right to left – in the example below, do the dilation first, then the reflection.

Example In the diagram below, the following composite transformation has been performed: $\triangle ABC \xrightarrow{d_{line\,\ell} \,\circ\, D_{P,\,0.5}} \triangle A''B''C''$

Discuss in detail the relationships between the three triangles giving specific reasons for your explanation.

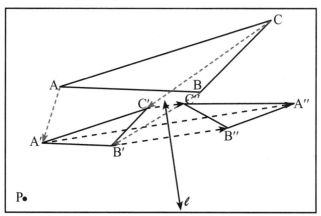

Solution: This is a composite transformation and the first step performed is the dilation about point P on triangle ABC to form its image $A'B'C'$. Dilations are non-rigid motions that preserve angle measure but not distance. The sides of the image have a ratio equal to the scale factor of the dilation. This makes $\triangle A'B'C' \sim \triangle ABC$.

Reflecting $\triangle A'B'C'$ over line ℓ, its image, $\triangle A''B''C''$ is formed. Reflections are rigid motion transformations, preserving distance and angle measure, making $\triangle A''B''C'' \sim \triangle A'B'C'$. Congruent figures are also similar.

Conclusion: Using the transitive property we can conclude:
$\triangle ABC \sim \triangle A'B'C'$ ***and*** $\triangle A'B'C' \sim \triangle A''B''C''$
$\therefore \triangle ABC \sim \triangle A''B''C''$

3.2

SIMILARITY AND CONGRUENCE OF TRIANGLES

Many types of problems can be developed that call for proving that two (or more) polygons are either similar or congruent to each other. (Remember that if figures are congruent, they are automatically similar as well.) In addition to the theorems, corollaries, axioms, definitions, properties, etc. that have already been discussed, there are other logical mathematical reasons that can be used to do a proof for congruence or similarity.

In this book the "given" and the "prove" (or goal) are described as "What is known?" and "What needs to be shown?" Some of the reasons available for use in developing the proof are listed without naming them as theorems, properties, etc. Follow the instructions of your teacher – some accept the wording of the reason, or the initials that describe the reason (SAS, CPCTC, etc.) while others prefer that the reason be formally named as a theorem, property, etc.

Note: Do not use the labels given in some textbooks like "Theorem 4-1" as a reason. Naming the reasons is not always consistent among textbooks.

CHARACTERISTICS OF TRIANGLES

Triangle: A triangle is a polygon with 3 sides.

Angles in a triangle: The sum of the angles in a triangle = 180°.

Types of triangles
- Classified by the measure of their angles.
    1. Acute: Each of the 3 angles is less than 90°.
    2. Right: One angle is 90°, the other two are a complementary pair.
    3. Obtuse: One angle is greater than 90° and less than 180°.
    4. *Equiangular: All 3 angles are congruent (60° each).

- Classified by the lengths of their sides.
    1. Scalene: None of the 3 sides are congruent.
    2. Isosceles: 2 sides are congruent.
    3. *Equilateral: All 3 sides are congruent.

    *Equilateral triangles are also equiangular. They are regular polygons.

Many different types of geometry problems involve the use of triangles. We are asked to use them in situations where they are congruent, similar, or they are parts of larger problems involving other polygons or circles. Many other applications are possible.

There are three methods used to prove that triangles are similar.

There are four methods used to prove that triangles are congruent and one for right triangles only.

General Triangle Information
Triangles and Angles

- The sum of the angles in any triangle is 180°.

  $m\angle 1 + m\angle 2 + m\angle 3 = 180$

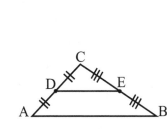

- An exterior angle of a triangle is equal to the sum of the two opposite (remote, nonadjacent) interior angles. $\angle 4 = \angle 2 + \angle 3$

- An exterior angle of a triangle is greater than either remote interior angle. $\angle 4 > \angle 2, \angle 4 > \angle 3$

- If two angles in a triangle are equal to the corresponding angles in another triangle, then the third angles in the triangles are congruent. This does not prove congruency of the triangles by itself. It must be combined with information about at least one pair of corresponding sides. It can be used to prove similarity of triangles.

Triangles and Sides

- The sum of any two sides of a triangle must be greater than the third side.

- Midsegments or Midlines: The segment joining the midpoints of two sides of a triangle is parallel to the third side and half as long as the third side.

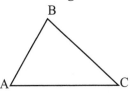

$$DE = \frac{1}{2}(AB)$$

- Base of a triangle
  1. Any triangle: When an altitude of a triangle is drawn, the side it is perpendicular to is called the "base" of the triangle.
  2. Isosceles triangle: The base of an isosceles triangle is the unequal side.
  3. Equilateral triangle: Any of the three sides can be considered the base. If an altitude is drawn, the side perpendicular to it is the base.

Relationships involving the sides and angles of a triangle:

- In a triangle, the side opposite the largest angle is the longest side. (The other two sides have the same relationship to their opposite angles. The middle sized side is opposite the middle sized angle and the shortest side is opposite the smallest angle.)

$\overline{AC}$ is the longest side

$\overline{BC}$ is the "middle sized" side

$\overline{AB}$ is the shortest side.

$\angle C < \angle A < \angle B$

Similarity and Congruence of Triangles

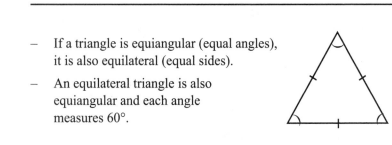

- If a triangle is equiangular (equal angles), it is also equilateral (equal sides).
- An equilateral triangle is also equiangular and each angle measures 60°.

- If two sides of a triangle are congruent, then the angles opposite those sides are congruent (isosceles triangle).
- If two angles of a triangle are congruent, the sides opposite them are congruent (isosceles).
- Base angles of an isosceles triangle are congruent. (Remember base angles are opposite the congruent sides.)

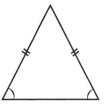

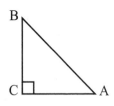

- In a right triangle, the hypotenuse, the side opposite the right angle, is always the longest side. $\overline{AB}$ is the longest side and is the hypotenuse.

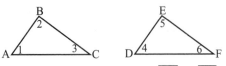

- Corresponding parts of congruent triangles are congruent. (Use for corresponding sides or corresponding angles in proofs.) [Often abbreviated as CPCTC.]

If $\triangle ABC \cong \triangle DEF$, then $\overline{AB} \cong \overline{DE}$, $\overline{BC} \cong \overline{EF}$, $\overline{AC} \cong \overline{DF}$ and $\angle 1 \cong \angle 4$, $\angle 2 \cong \angle 5$, $\angle 3 \cong \angle 6$

Geometry Made Easy – Common Core Standards Edition

ISOSCELES TRIANGLES

This information is included with the general triangle information, but it is specific to isosceles triangles.

- **Known:** A triangle is isosceles (given, or already proven)
    - In an isosceles triangle, the angles opposite the congruent sides (base angles) are congruent.
    - In an isosceles triangle, two sides are congruent.
    - The altitude of an isosceles triangle drawn from the base divides the triangle into two congruent right triangles.

$\triangle ABD \cong \triangle CBD$

Note: The altitude of an isosceles triangle can often be found using the Pythagorean Theorem. See page 76 for Isosceles triangles and the Pythagorean Theorem.

RIGHT TRIANGLES
(See also page 75)

Hypotenuse - Leg Theorem: In two right triangles, if the hypotenuse and a leg of one are congruent to the corresponding parts of the other, the triangles are congruent. It must be stated that the triangles are right triangles first.

$\triangle ABC \cong \triangle DEF$

Additional Right Triangle Information:
The midpoint of the hypotenuse of a right triangle is equidistant from the three vertices. The two acute angles in a right triangle are complementary.

Pythagorean Theorem: c is the *hypotenuse* and a and b are the legs. Side c is the longest side. If a triangle is a right triangle, then the square of the length of the hypotenuse is equal to the sum of the squares of the lengths of the legs. As a formula it is written: $c^2 = a^2 + b^2$. It is also commonly written as $a^2 + b^2 = c^2$.

Note: The Pythagorean Theorem is used to find a missing side of a triangle, and it also can be used to prove whether or not a triangle is a right triangle. If the sides of a given triangle can be substituted into the Pythagorean Theorem, using the longest side as c and the numbers check, then the triangle is a right triangle.

Similarity and Congruence of Triangles

Theorem: In a right triangle, the sum of the squares of the legs is equal to the square of the hypotenuse.

Examples

❶ **Given:** Figure 1, right triangle *ABC,* with hypotenuse *c*, legs *a* and *b*.

Prove: $a^2 + b^2 = c^2$

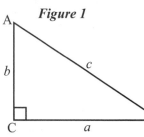

Figure 1

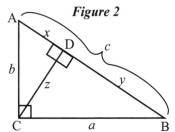

Figure 2

Statement	Reason
1. Right △*ABC*	1. Given.
2. Draw altitude from right angle *C* perpendicular to $\overline{AB}$ at *D*. Label as shown in Figure 2.	2. An altitude of a triangle is drawn from a vertex of a triangle and is perpendicular to the side opposite the vertex. (Labels are for convenience.)
3. ∠*ADC* ≅ ∠*BDC*	3. Perpendicular lines form right angles and all right angles are congruent.
4. ∠*B* ≅ ∠*B* ; ∠*A* ≅ ∠*A*	4. Reflexive Property.
5. ∠*BDC* ≅ ∠*BCA* ; ∠*ADC* ≅ ∠*BCA*	5. All right angles are congruent.
6. △*BDC* ~ △*BCA* ; △*ADC* ~ △*BCA*	6. If two angles of a triangle are congruent to two angles in another triangle, the triangles are similar. AA similarity criterion.
7. In △*BDC* & △*BCA* : $\dfrac{a}{c} = \dfrac{y}{a}$ In △*ADC* & △*BCA* : $\dfrac{b}{c} = \dfrac{x}{b}$	7. In similar triangles, corresponding sides are proportional.
8. $a^2 = cy$; $b^2 = cx$	8. In a proportion, the product of the means equals the product of the extremes.
9. $y + x = c$	9. Segment addition.
10. $a^2 + b^2 = cy + cx$	10. Addition property of equality.
11. $a^2 + b^2 = c(y + x)$	11. Factor the right side of the equation.
12. $a^2 + b^2 = c^2$	12. Substitution.

Note: There are many proofs of this theorem.

Geometry Made Easy – Common Core Standards Edition

Alternate method to prove the Pythagorean Theorem using similarity. This is a less formal proof than the statement-reason proof. In it, the technique of redrawing parts of a diagram separately is used, so they are more easily identified.

❷ **Given:** Right triangle *ABC* with altitude *CD* drawn to the hypotenuse.

Prove: Using similarity, prove that the sum of the legs squared is equal to the square of the hypotenuse.

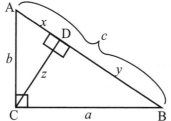

Paragraph Proof: Draw the three triangles separately and label appropriately. Each of the smaller right triangles shares one angle with the large triangle making each one similar to the large triangle, *ABC*, using AA similarity. Using the transitive property, they are similar to each other. Since they are similar, their sides are proportional.

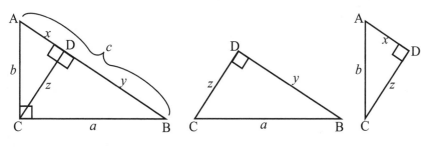

Triangles *ABC* & *ACD*

$$\frac{AC}{AB} = \frac{AD}{AC}$$

$$(AC)^2 = AB \cdot AD$$

Triangles *ABC* & *CBD*

$$\frac{BC}{AB} = \frac{DB}{BC}$$

$$(BC)^2 = DB \cdot AB$$

$$(AC)^2 + (BC)^2 = AB \cdot AD + DB \cdot AB$$
$$(AC)^2 + (BC)^2 = AB(AD + DB)$$
$$(AC)^2 + (BC)^2 = AB(AB)$$
$$(AC)^2 + (BC)^2 = (AB)^2$$

Conclusion: The sum of the squares of the legs of a right triangle is equal to the square of the hypotenuse.

Similarity and Congruence of Triangles

TRIANGLE SIMILARITY THEOREMS

Using the information we have about rigid and non-rigid transformations, along with properties, definitions, and theorems, including corollaries and axioms, we can prove theorems about triangle similarity.

Examples

❶ **Theorem:** A line parallel to one side of a triangle divides the other two sides proportionally.

Given: $\triangle KLM$ with $\overline{PQ} \parallel \overline{LM}$ which forms $\triangle KPQ$

Prove: $\dfrac{PL}{KP} = \dfrac{QM}{KQ}$

Plan: Since proportionality is part of this problem, prove that the two triangles are similar before proceeding to the ratios of the sides.

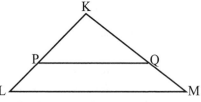

Statement	Reason
1. $\overline{PQ} \parallel \overline{LM}$	1. Given.
2. $\angle KPQ \cong \angle KLM$; $\angle KQP \cong \angle KML$	2. When parallel lines are cut by a transversal, corresponding angles are congruent.
3. $\angle K \cong \angle K$	3. Reflexive Property.
4. $\triangle KLM \sim \triangle KPQ$	4. If 3 angles of a triangle are congruent to the corresponding angles of another triangle, the triangles are similar. AAA Similarity.
5. $\dfrac{KL}{KP} = \dfrac{KM}{KQ}$	5. Corresponding sides of similar triangles are proportional.
6. $KP + PL = KL$; $KQ + QM = KM$	6. The whole is equal to the sum of its parts.
7. $\dfrac{KP+PL}{KP} = \dfrac{KQ+QM}{KQ}$	7. Substitution.
8. $1 + \dfrac{PL}{KP} = 1 + \dfrac{QM}{KQ}$	8. Simplification of fractions.
9. $\dfrac{PL}{KP} = \dfrac{QM}{KQ}$	9. Subtract 1 from both sides of the equation.

Similarity and Congruence of Triangles

❷ Theorem: If a line divides two sides of a triangle proportionally, it is parallel to the third side.

Given: $\triangle ACE$ with $\overline{BD}$ which forms $\triangle BCD$.

$$\frac{AB}{BC} = \frac{ED}{DC}$$

Prove: $\overline{BD} \parallel \overline{AE}$

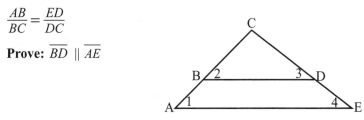

Plan: Prove the triangles are similar, then use the congruency of the corresponding angles to prove the parallelism.

(*Note:* $\overline{AC}$ and $\overline{EC}$ intersect both lines and are transversals.)

Paragraph Proof: Given triangles ACE and BCD with segment BD and $\frac{AB}{BC} = \frac{ED}{DC}$. Angle C is contained in both triangles and is equal to itself using the reflexive property. Use addition equality to add one in the form of $\frac{BC}{BC}$ and $\frac{CD}{CD}$ and on the left and right sides of the equation respectively, the equation is then $\frac{BC}{BC} + \frac{AB}{BC} = \frac{DC}{DC} + \frac{ED}{DC}$. $AB + BC = AC$ and $ED + CD = EC$ using segment addition. Since there is a common denominator on each side of the equation, it is easy to combine the fractions to $\frac{BC + AB}{BC} = \frac{DC + ED}{DC}$. Now, using substitution, replace the numerator of each fraction with their equivalent terms and determine that $\frac{AC}{BC} = \frac{EC}{DC}$. This makes $\triangle ACE \sim \triangle BCD$ using the SAS similarity criterion which says that if two sides of a triangle are proportional to the corresponding sides of another and the angle included between those sides is congruent in both triangles, the triangles are similar. Since corresponding angles in similar triangles are congruent, $\angle 1 \cong \angle 2$ and $\angle 3 \cong \angle 4$. In conclusion, $\overline{BD} \parallel \overline{AE}$ because if two lines are cut by a transversal and corresponding angles are congruent, the lines are parallel.

TRIANGLE SIMILARITY

In proving that triangles are similar, we can use three different methods.

1. **AA:** Prove that two of the three angles in one triangle are congruent to two angles in the other. If two angle of a triangle are congruent or equal to two angles of another, the third angle in each must be congruent as well, since the sum of the angles in each triangle is 180°. AA is equivalent to AAA.

2. **SSS:** Prove that each side of one triangle is proportionate to the corresponding side of the other triangle.

3. **SAS:** Prove that two sides of one triangle are proportionate to the two corresponding sides of the other and that the included angle is congruent in both triangles.

Examples Triangle Similarity Proofs

❶ **Statement Reason Proof**
   Given: $\triangle AFD$. Parallelogram $ABCD$.
   Point B is on side $\overline{AF}$ and E is on $\overline{BC}$.
   Prove: $\triangle BEF \sim \triangle ADF$

Statement	Reason
1. $\triangle AFD$, Parallelogram $ABCD$	1. Given.
2. $\overline{AD} \parallel \overline{BC}$	2. Opposite sides of a parallelogram are parallel.
3. $\angle FBE \cong \angle FAD$ $\angle FEB \cong \angle FDA$	3. When 2 $\parallel$ lines are cut by a transversal, corresponding angles are $\cong$.
4. $\triangle BEF \sim \triangle ADF$	5. AA $\cong$ AA

(Examples 1 and 2 have the same diagram, but different triangles are proven similar.)

❷ **Paragraph Proof**

Given: Parallelogram $ABCD$.
$\overline{AB}$ is extended to F and $\overline{FD}$ is
drawn. $\overline{FD}$ intersects $\overline{BC}$ at E.

Prove: $\triangle BEF \sim \triangle CED$

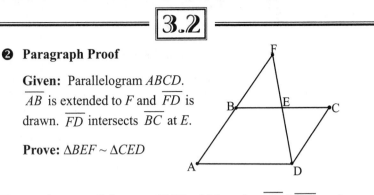

We are given parallelogram $ABCD$ which makes $\overline{AB} \parallel \overline{DC}$ and
$\overline{BC} \parallel \overline{AD}$ by definition. $\angle FBE \cong \angle DCE$ and $\angle BFE \cong \angle CDE$ because
when parallel lines are cut by a transversal, alternate interior angles are
congruent. $\angle FEB \cong \angle DEC$ because vertical angles are congruent.
Therefore $\triangle BEF \sim \triangle CED$ because when two angles in a triangle are
congruent to two angles in another triangle, the triangles are similar.

❸ **Given:** $\triangle ABC$. E is the midpoint of $\overline{AB}$,
D is the midpoint of $\overline{BC}$.

Prove: $\triangle ABC \sim \triangle EBD$

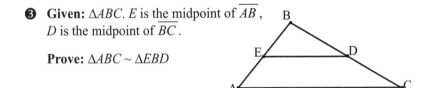

Statement	Reasons
1. $\triangle ABC$, D is the midpoint of $\overline{BC}$, E is the midpoint of $\overline{AB}$	1. Given.
2. $\overline{ED} \parallel \overline{AC}$ and $\dfrac{ED}{AC} = \dfrac{1}{2}$	2. A segment joining the midpoints of 2 sides of a triangle is parallel to the 3rd side and equal to $\dfrac{1}{2}$ of its measure.
3. $BD = \dfrac{1}{2}(BC)$, $BE = \dfrac{1}{2}(AB)$	3. A midpoint divides a segment in half.
4. $\triangle ABC \sim \triangle EBD$	4. Two triangles are similar if 3 sides of one are proportional to the corresponding sides of the other.

Note: This proof could be done using AA as well. Since $\angle B$ is shared by
both triangles, it equals itself by the reflexive property. Since $\overline{ED} \parallel \overline{AC}$,
$\angle BED \cong \angle BAC$ and $\angle BDE \cong \angle BCA$ because they are pairs of
corresponding angles which are congruent. That makes the two
triangles similar by AA.

Examples **Using calculations in Similar Triangles**

❶ Two triangles are similar. The sides of one triangle measure 23, 27, and 40. The middle sized side of the second triangle is 81. Find the lengths of the shortest side and longest side of the second triangle.

Solution: Find the ratio of the measures of the middle sides – they are the corresponding sides whose measures are given in the two triangles. This is also called the constant of proportionality.

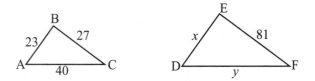

$$\triangle ABC \sim \triangle DEF$$

Use this ratio to find the values of the other sides by making proportions.

$$\frac{27}{81} = \frac{1}{3}$$

$$\frac{23}{x} = \frac{1}{3} \quad \text{and} \quad \frac{40}{y} = \frac{1}{3}$$

$$x = 69 \qquad y = 120$$

The shortest side of the second triangle is 69 and the longest side is 120.

Similarity and Congruence of Triangles

RIGHT TRIANGLES

Pythagorean Theorem $a^2 + b^2 = c^2$

Examples

❶ **Known:** Measures of two sides of a right triangle.
Show: Measure of the third side.

The hypotenuse of a right Δ is 17.
One leg is 8. Find the length of
the other leg.

$8^2 + b^2 = 17^2$

$64 + b^2 = 289$

$b^2 = 225$

$b = 15$

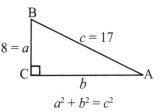

❷ **Known:** Measure of three sides of a triangle.
Show: The triangle is or is not a right triangle.

– In a triangle, if the square of the longest side equals the sum of the
squares of the other two sides, the triangle is a right triangle.

Is a triangle with sides of 5, 10, and $5\sqrt{3}$ a right triangle?

Solution: If the sides "check" in the Pythagorean Theorem, it is a
right triangle. The longest side is the hypotenuse. ($5\sqrt{3} \approx 8.66$).

$\left(5\sqrt{3}\right)^2 + (5)^2 \stackrel{?}{=} 10^2$

$75 + 25 \stackrel{?}{=} 100$

$100 = 100 \; \sqrt{}$

This is a right triangle.

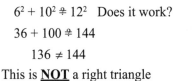

❸ Is a triangle with sides of 6, 10, and 12 a right triangle?

Solution: Only sides that "check" in the Pythagorean Theorem can
form right triangles. The hypotenuse is always the longest side.

$6^2 + 10^2 \stackrel{?}{=} 12^2$ Does it work?

$36 + 100 \stackrel{?}{=} 144$

$136 \neq 144$

This is **NOT** a right triangle

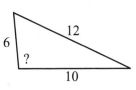

Similarity and Congruence of Triangles

Isosceles Triangles and The Pythagorean Theorem: Since the altitude of an isosceles triangle is the perpendicular bisector of the base of the triangle, the Pythagorean Theorem is often used in calculations for isosceles (or equilateral) triangles.

Example

Find the height of isosceles triangle ABC. The base, $\overline{AC}$, is 10 inches and the congruent sides are each 14 inches.

Solution: The altitude divides the base into 2 equal segments, each 5 inches. The 14 inch side becomes the hypotenuse formed when the altitude, h, is drawn.

$a^2 + b^2 = c^2$
$5^2 + b^2 = 14^2$
$b^2 = 196 - 25$
$b = \sqrt{171}$ inches
$\therefore h = \sqrt{171}$ inches

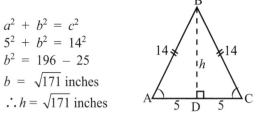

Pythagorean Triples: There are some common right triangle measurements or multiples of them that are helpful to recognize. They are called Pythagorean Triples. Examples: 3,4,5; 5,12,13; and 8,15,17. Remember the longest side is always the hypotenuse. There are many more Pythagorean Triples, but these are the most common.

Proportions in Right Triangles

Altitude Drawn to the Hypotenuse of a Right Triangle:

1) Similar triangles are formed

When the altitude of a right triangle is drawn to the hypotenuse, two similar triangles are created. There are some theorems that can be applied directly to this situation, but my students have found it helpful to see the similarity of the triangles as well.

In triangle ABC, C is the right angle. The altitude, $\overline{CD}$ is drawn to hypotenuse $\overline{AB}$. The diagram shows the relationship between the three triangles – they are taken apart to show them more clearly. The similarity relationship is true no matter what the acute angle measures are in the original right triangle. The example on the next page has angles of 90°, 25°, and 65° to demonstrate.

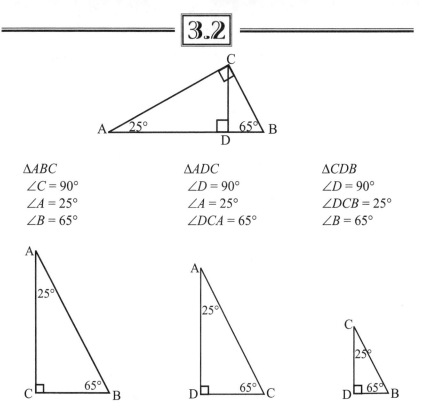

$\triangle ABC$	$\triangle ADC$	$\triangle CDB$
$\angle C = 90°$	$\angle D = 90°$	$\angle D = 90°$
$\angle A = 25°$	$\angle A = 25°$	$\angle DCB = 25°$
$\angle B = 65°$	$\angle DCA = 65°$	$\angle B = 65°$

Since all three triangles have corresponding angles that are congruent, the triangles are all similar.

2) Mean Proportionals in Right Triangles with an Altitude:

Known: An altitude is drawn to the hypotenuse of a right triangle.

Show: Relationships between segments of a right triangle.
Find length of segments.

a) The altitude drawn to the hypotenuse of a right triangle divides the triangle into two triangles which are similar (~) to each other and similar to the original triangle. $\triangle ADC \sim \triangle CDB \sim \triangle ACB$.

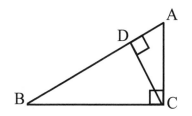

Similarity and Congruence of Triangles

b) In a right triangle with the altitude drawn to the hypotenuse, the length of each leg is the mean proportional between the length of the segment of the hypotenuse that is adjacent to the leg and the whole length of the hypotenuse. $\dfrac{AD}{AC} = \dfrac{AC}{AB}$ and $\dfrac{BD}{BC} = \dfrac{BC}{AB}$

Note: The part of the hypotenuse that is adjacent to a leg is sometimes called, "The shadow of the leg." In the example below $\overline{BD}$ is the shadow of $\overline{BC}$ and $\overline{AD}$ is the shadow of $\overline{AC}$.

Example **Right triangles with an altitude**

In $\triangle ABC$, CD is the altitude drawn to hypotenuse $\overline{AB}$. $BC = 10$, $BD = 4$. Find the length of AB.

Solution: Leg BC is the mean proportional between the segment of the hypotenuse it is adjacent to and the entire hypotenuse.

Let $x = AD$, $x + 4 = AB$

$$\frac{4}{10} = \frac{10}{x + 4}$$
$$4x + 16 = 100$$
$$4x = 84$$
$$x = 21, \ AD = 21$$
$$x + 4 = 21 + 4 = 25$$
$$AB = 25$$

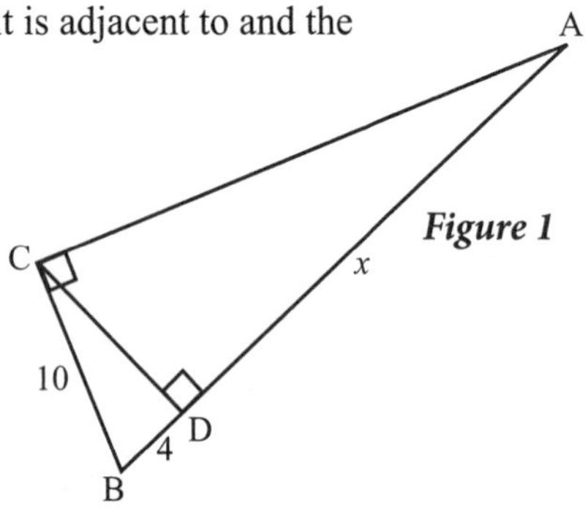

Figure 1

c) In a right triangle with the altitude drawn to the hypotenuse, the length of the altitude is the mean proportional between the lengths of the segments of the hypotenuse. $\dfrac{AD}{CD} = \dfrac{CD}{DB}$

Example If $CD = 6$ and BD is 9 less than AD, find AD and DB.

Solution: The altitude is the mean proportional between the segments it forms on the hypotenuse.

Let x = length of AD

$$\frac{AD}{CD} = \frac{CD}{DB}$$
$$\frac{x - 9}{6} = \frac{6}{x}$$
$$x(x - 9) = 36$$
$$x^2 - 9x = 36$$
$$x^2 - 9x - 36 = 0$$
$$x = 12 \quad x = -3 \ \text{reject}$$
$$x - 9 = 12 - 9 = 3$$
$$AD = 12, \ BD = 3$$

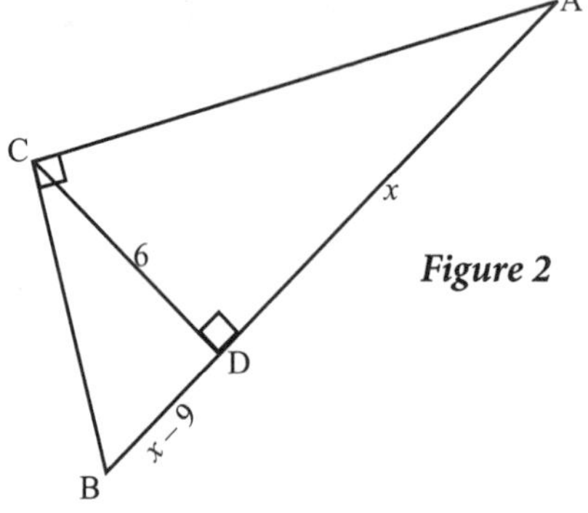

Figure 2

If two triangles are congruent, they are "exact matches" of each other. One could be placed on top of the other and the corresponding sides would be the same length and the corresponding angles would have the same measure. "Included" means it is located between the other two parts described.

Note: See also Rigid Motion Transformations in Unit 2.

1) **Show:** Triangles are congruent. Use any of the following reasons.

- **ASA ≅ ASA** (angle-side-angle)
  If two angles and the included side in one triangle are congruent to the corresponding parts of the other, the triangles are congruent.

- **SAS ≅ SAS** (side-angle-side)
  If two sides and the included angle in one triangle are congruent to the corresponding parts of the other, the triangles are congruent.

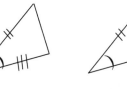

- **SSS ≅ SSS** (side-side-side)
  If three sides of one triangle are congruent to three sides of another the two triangles are congruent.

- **AAS ≅ AAS** (angle-angle-side)
  If two angles and a non-included side of one triangle are congruent to the corresponding parts of another, the triangles are congruent.

- **HL** (Hypotenuse - Leg Theorem)
  If the hypotenuse and one leg of a right triangle are congruent to the corresponding parts of the other, the two triangles are congruent. Triangles must be identified as right triangles before using HL.

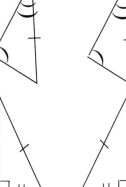

Note: SSA ≅ SSA is **not** an adequate proof for triangle congruency due to the "Ambiguous Case" which will be studied in Trigonometry.

Similarity and Congruence of Triangles

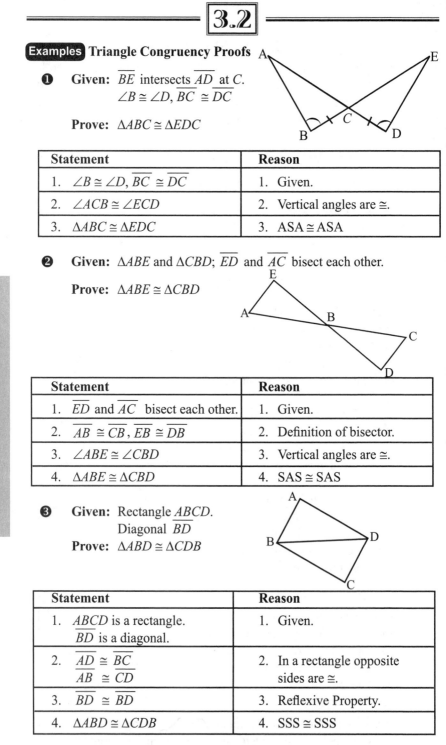

$$\boxed{3.2}$$

Examples **Triangle Congruency Proofs**

❶ **Given:** $\overline{BE}$ intersects $\overline{AD}$ at C.
$\angle B \cong \angle D$, $\overline{BC} \cong \overline{DC}$

Prove: $\triangle ABC \cong \triangle EDC$

Statement	Reason
1. $\angle B \cong \angle D$, $\overline{BC} \cong \overline{DC}$	1. Given.
2. $\angle ACB \cong \angle ECD$	2. Vertical angles are $\cong$.
3. $\triangle ABC \cong \triangle EDC$	3. ASA $\cong$ ASA

❷ **Given:** $\triangle ABE$ and $\triangle CBD$; $\overline{ED}$ and $\overline{AC}$ bisect each other.

Prove: $\triangle ABE \cong \triangle CBD$

Statement	Reason
1. $\overline{ED}$ and $\overline{AC}$ bisect each other.	1. Given.
2. $\overline{AB} \cong \overline{CB}$, $\overline{EB} \cong \overline{DB}$	2. Definition of bisector.
3. $\angle ABE \cong \angle CBD$	3. Vertical angles are $\cong$.
4. $\triangle ABE \cong \triangle CBD$	4. SAS $\cong$ SAS

❸ **Given:** Rectangle $ABCD$.
Diagonal $\overline{BD}$
Prove: $\triangle ABD \cong \triangle CDB$

Statement	Reason
1. $ABCD$ is a rectangle. $\overline{BD}$ is a diagonal.	1. Given.
2. $\overline{AD} \cong \overline{BC}$ $\overline{AB} \cong \overline{CD}$	2. In a rectangle opposite sides are $\cong$.
3. $\overline{BD} \cong \overline{BD}$	3. Reflexive Property.
4. $\triangle ABD \cong \triangle CDB$	4. SSS $\cong$ SSS

Similarity and Congruence of Triangles

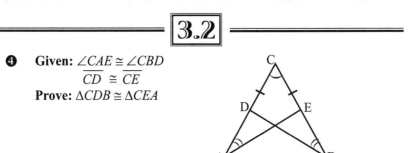

4 **Given:** $\angle CAE \cong \angle CBD$
$\overline{CD} \cong \overline{CE}$
Prove: $\triangle CDB \cong \triangle CEA$

Paragraph Proof:

$\angle CAE \cong \angle CBD$ and $\overline{CD} \cong \overline{CE}$ because this is the given information. $\angle ACE \cong \angle BCD$ by the reflexive property. The two triangles, $\triangle CDB$ and $\triangle CEA$ are congruent because one triangle has two angles and one side congruent to the corresponding parts of the other.

Therefore: $\triangle CDB \cong \triangle CEA$

Note: When overlapping triangles are involved, it often helps to use a different colored pencil for each triangle you are working, with in order to see "what's what" more easily!

5 **Given:** $\overline{AC} \perp \overline{DB}$
$\overline{AD} \cong \overline{AB}$
Prove: $\triangle DCA \cong \triangle BCA$

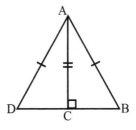

Statement	Reason
1. $\overline{AC} \perp \overline{DB}$, $\overline{AD} \cong \overline{AB}$	1. Given.
2. $\angle ACD$ and $\angle ACB$ are right angles.	2. $\perp$ lines form right angles.
*3. $\triangle DCA$ and $\triangle BCA$ are right triangles.	3. Definition of right triangle.
4. $\overline{AC} \cong \overline{AC}$	4. Reflexive Property.
5. $\triangle DCA \cong \triangle BCA$	5. HL $\cong$ HL

* The triangles must be proven and stated to be to be right triangles before using HL $\cong$ HL.

Similarity and Congruence of Triangles

❻ **Given:** $\overline{AB} \cong \overline{AE}$
$\overline{BC} \cong \overline{DE}$

Prove: $\triangle ACD$ is isosceles.

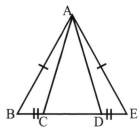

Flow Proof

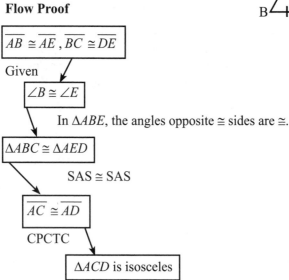

$\overline{AB} \cong \overline{AE}$, $\overline{BC} \cong \overline{DE}$

Given

$\angle B \cong \angle E$

In $\triangle ABE$, the angles opposite $\cong$ sides are $\cong$.

$\triangle ABC \cong \triangle AED$

SAS $\cong$ SAS

$\overline{AC} \cong \overline{AD}$

CPCTC

$\triangle ACD$ is isosceles

A triangle with two $\cong$ sides is isosceles.

SIMILARITY AND CONGRUENCE OF POLYGONS

POLYGONS IN GENERAL

Polygon: A plane figure in which all sides are line segments. A polygon is named by the number of sides it contains. In general, a polygon with "*n*" sides is called an "*n*-gon". (Ex: A hexagon can also be called a 6-gon).

We will work primarily with *convex polygons* in which all the vertices point outward - think of the shape of a stop sign.

In a *concave* polygon, at least one pair of sides points inward making a cave like appearance to that side - think of a star shaped polygon.

Known: A specific *n*-gon is described. It is convex.

Show: The measures of its **interior** and/or **exterior** angles.

- The formula **S** = (*n* − 2)(180°) gives the sum of the **interior** angles of any convex polygon with *n* sides.
  $\angle 1 + \angle 2 + \angle 3 + \angle 4 + \angle 5$
  $= (5 - 2)(180) = 540°$

- The sum of the **exterior** angles of any convex polygon is 360°.
  $\angle a + \angle b + \angle c + \angle d + \angle e = 360°$

- An interior angle and its adjacent exterior angle are supplementary angles: $\angle 1 + \angle a = 180°$. If information is given that enables you to know the measure of an exterior angle, you can find the interior angle by subtracting from 180 degrees. Use the same process if you know the measure of the interior angle to find the adjacent exterior angle.

Regular Polygon: A polygon in which all sides are congruent *and* all angles are congruent.

Known: A regular polygon is either given or described and it has *n* sides.

Show: The value of its interior and/or exterior angles.

- The sum of the exterior angles in any convex polygon is 360°. Since the exterior angles of a regular polygon are congruent, 360/*n* will give the measure of an exterior angle of an *n*-sided regular polygon.

- The sum of the interior angles of any convex polygon is (*n* − 2)(180) and since all the angles are equal in a regular polygon, to find the measure of one interior angle, divide by *n*.

Similarity and Congruence of Polygons

Angles of Polygon Examples:

Examples

❶ Find the measure of an interior angle of an 8 sided regular polygon (octagon). $n = 8$

Formula: $\dfrac{(n-2)180}{n}$ An interior $\angle$ of a regular octagon measure $135°$.

$\dfrac{(8-2)(180)}{8} = 135°$

❷ The exterior angle of a regular n-gon measures $24°$. Find the value of n.

Solution: The sum of the exterior angles of any n-gon $= 360°$. Divide $360°$ by the measure of one exterior angle in a regular polygon to find the number of sides.

$\dfrac{360}{24} = n$

$n = 15$

Congruent Polygons: Two (or more) polygons that have corresponding sides that are congruent and corresponding angles that are congruent.

❸ A game requires a playing field the shape of a regular pentagon. At Baker Park, a playing field is built with each side measuring 20 feet. The town wants to build another playing field to match the one in Baker Park at Cook Park. To build the playing field in Cook Park, what measurements must be used for its construction?

Solution: All regular pentagons have 5 congruent angles and 5 congruent sides. The measure of each interior angle is $\dfrac{(n-2)(180)}{n} = \dfrac{(5-2)180}{5} = 108°$. In Baker Park the length of each side is 20 feet. In order to make a new field in Cook Park, each interior angle must be $108°$ and each side must be 20 feet in length. The two playing fields will be congruent since they will have corresponding angles that are congruent and corresponding sides that are also congruent.

Similarity and Congruence of Polygons

SIMILAR POLYGONS AND PROPORTIONS

- If two polygons are similar, each angle in one is congruent to the corresponding angle of the other, and each side in one is proportional to the corresponding side of the other.

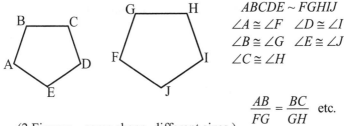

$$ABCDE \sim FGHIJ$$
$$\angle A \cong \angle F \quad \angle D \cong \angle I$$
$$\angle B \cong \angle G \quad \angle E \cong \angle J$$
$$\angle C \cong \angle H$$

(2 Figures – same shape, different sizes.)

$$\frac{AB}{FG} = \frac{BC}{GH} \text{ etc.}$$

- The scale factor or ratio factor of two similar polygons is the ratio of the side of one to the corresponding side of the other.

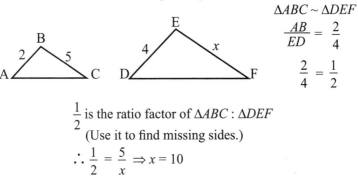

$$\triangle ABC \sim \triangle DEF$$
$$\frac{AB}{ED} = \frac{2}{4}$$
$$\frac{2}{4} = \frac{1}{2}$$

$\frac{1}{2}$ is the ratio factor of $\triangle ABC : \triangle DEF$
(Use it to find missing sides.)

$$\therefore \frac{1}{2} = \frac{5}{x} \Rightarrow x = 10$$

- If two polygons are similar and have a scale factor of $a : b$, the ratio of their perimeters is $a : b$.

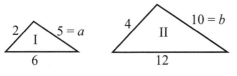

$$\triangle I \sim \triangle II$$
Ratio of $\frac{a}{b} = \frac{5}{10} = \frac{1}{2}$

Perimeter $\triangle I = 13$
Perimeter $\triangle II = 26$

$$\frac{P_I}{P_{II}} = \frac{13}{26} = \frac{1}{2}$$

Similarity and Congruence of Polygons

- If two polygons are similar and have a scale factor of $a : b$, the ratio of their areas is $a^2 : b^2$.

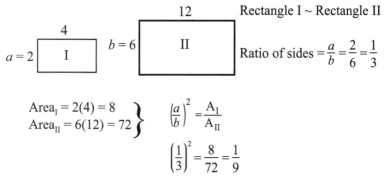

Rectangle I ~ Rectangle II

Ratio of sides $= \dfrac{a}{b} = \dfrac{2}{6} = \dfrac{1}{3}$

$\left. \begin{array}{l} \text{Area}_I = 2(4) = 8 \\ \text{Area}_{II} = 6(12) = 72 \end{array} \right\} \quad \left(\dfrac{a}{b}\right)^2 = \dfrac{A_I}{A_{II}}$

$$\left(\dfrac{1}{3}\right)^2 = \dfrac{8}{72} = \dfrac{1}{9}$$

- If a line intersects the sides a triangle and is parallel to the third side, then the line divides the two intersected sides proportionally.

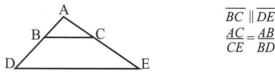

$\overline{BC} \parallel \overline{DE}$
$\dfrac{AC}{CE} = \dfrac{AB}{BD}$

- If three parallel lines intersect two transversals, they divide the transversals proportionally.

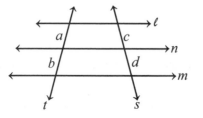

$\ell \parallel n \parallel m$
t and s are transversals.
$\dfrac{a}{b} = \dfrac{c}{d}$

Notice that the parallel lines and transversals create two similar trapezoids.

QUADRILATERALS

Quadrilaterals - 4 sided polygons: The relationship between the various kinds of quadrilaterals is shown here. The characteristics of each quadrilateral are used in many types of geometry problems and proofs.

MEMORIZE THEM!

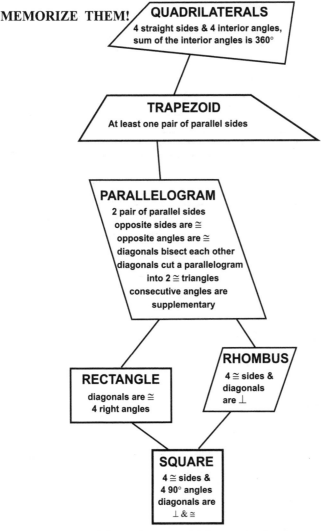

QUADRILATERALS
4 straight sides & 4 interior angles,
sum of the interior angles is 360°

TRAPEZOID
At least one pair of parallel sides

PARALLELOGRAM
2 pair of parallel sides
opposite sides are ≅
opposite angles are ≅
diagonals bisect each other
diagonals cut a parallelogram
into 2 ≅ triangles
consecutive angles are
supplementary

RECTANGLE
diagonals are ≅
4 right angles

RHOMBUS
4 ≅ sides &
diagonals
are ⊥

SQUARE
4 ≅ sides &
4 90° angles
diagonals are
⊥ & ≅

Similarity and Congruence of Polygons

Note: The definition of trapezoid is somewhat controversial. The "inclusive" definition is the one used here. It says that a trapezoid has at least one pair of parallel sides. The "exclusive" definition says a trapezoid has exactly one pair of parallel sides. It excludes quadrilaterals with both pair of opposite sides that are parallel.

TRAPEZOIDS

Trapezoids: A subset of quadrilaterals, have *at least one pair* of parallel sides.

Known: A quadrilateral is given.

Show: It is a **trapezoid** by proving
– One pair of sides is parallel.

Examples **TRAPEZOID Examples**

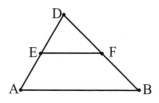

❶ Given: $\triangle DAB$, $\overline{EF}$ is drawn
$\triangle DEF \sim \triangle DAB$

Prove: $ABFE$ is a trapezoid.

Statement	Reason
1. $\triangle DAB$, $\overline{EF}$ is drawn. $\triangle DEF \sim \triangle DAB$	1. Given.
2. $\angle DEF \cong \angle DAB$	2. In similar triangles, corresponding angles are congruent.
3. $\overline{EF} \parallel \overline{AB}$	3. When 2 lines are cut by a transversal and have corresponding angles that are congruent, the lines are parallel.
4. $ABFE$ is a trapezoid.	4. A quadrilateral with at least one pair of parallel sides is a trapezoid.

❷ In trapezoid $KJLM$, $\angle MKJ \cong \angle LJK$ and $\angle KML = \text{m}\angle JLM$.

$m\angle KML = 2(2x + 10)$
and $m\angle JLM = 5(x - 1)$

Find the measures of each of the 4 angles.

Solution: Set the values of the two congruent angles equal to each other. Find the measure of those angles.

$2(2x + 10) = 5(x - 1)$
$4x + 20 = 5x - 5$
$25 = x$
$\angle KML = \angle JLM = 5(25 - 1) = 120°$

Since: $\overline{KJ} \parallel \overline{ML}$, $m\angle MKJ + m\angle KML = 180°$;
and $m\angle LJK = m\angle MKJ = (180 - 120)° = 60°$.

PARALLELOGRAMS

This section provides "reasons" that can be applied to various kinds of quadrilaterals. We will work with two subsets of quadrilaterals - parallelograms and trapezoids.

A parallelogram is a subset of the set of quadrilaterals. It has many characteristics that can be used to prove the figure is a parallelogram. If the quadrilateral is already known to be a parallelogram, these facts can be used to develop further geometric relationships in the subsets of parallelograms which include rectangles, rhombuses (sometimes the plural is written as rhombi), and squares. In order to work with any of the subsets of parallelograms, it is necessary to *prove that the figure is, first of all, a parallelogram.*

Known: A quadrilateral is given.
Show: It is a **parallelogram** by proving any one of the following:
– one pair of opposite sides are both parallel and congruent.
– both pairs of opposite sides are parallel.
– both pairs of opposite sides congruent.
– both pairs of opposite angles are congruent.
– the diagonals of the quadrilateral bisect each other.

Note: A quadrilateral must be proven to be a parallelogram before continuing with work involving the subsets.

Similarity and Congruence of Polygons

Examples Prove a quadrilateral is a parallelogram.

❶ Given: Parallelogram *BHDG*, $\overline{AH} \cong \overline{GC}$
Prove: *ABCD* is a parallelogram.

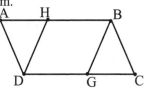

Statement	Reason
1. Parallelogram *BHDG*, $\overline{AH} \cong \overline{GC}$	1. Given.
2. $\overline{BH} \cong \overline{DG}$, $\overline{BH} \parallel \overline{DG}$ (making $\overline{AB} \parallel \overline{DC}$)	2. Opposite sides of a parallelogram are congruent and parallel.
3. $\overline{BH} + \overline{HA} \cong \overline{DG} + \overline{GC}$	3. Addition Axiom.
4. $\overline{BH} + \overline{HA} \cong \overline{AB}$ $\overline{DG} + \overline{GC} \cong \overline{CD}$	4. Segment Addition.
5. $\overline{AB} \cong \overline{CD}$	5. Substitution.
6. *ABCD* is a parallelogram.	6. If a quadrilateral has one pair of opposite sides that are both parallel and congruent, it is a parallelogram.

❷ Given: $\triangle AEB \cong \triangle CED$
Prove: $\triangle ABCD$ is a parallelogram.

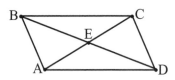

Statement	Reason
1. $\triangle AEB \cong \triangle CED$	1. Given.
2. $\angle DBA \cong \angle BDC$, $\overline{AB} \cong \overline{CD}$	2. CPCTC
3. $\overline{AB} \parallel \overline{CD}$	3. If two lines are cut by a transversal and alternate interior angles are $\cong$, the lines are parallel.
4. *ABCD* is a parallelogram.	4. If one pair of sides in a quadrilateral is both parallel and congruent, the figure is a parallelogram.

Note: There are many other ways to do this proof. One other method would be to use CPCTC to show that the diagonals bisect each other, if so, then that too would be proof of a parallelogram.

RHOMBUS

Show: A quadrilateral is a **rhombus**. (Again, it has everything a parallelogram has, and specific proof is needed).

- A parallelogram with two adjacent sides that are congruent is a rhombus.
- A parallelogram with perpendicular diagonals is a rhombus.
- A parallelogram with diagonals that bisect the angles of the parallelogram is a rhombus.

Examples Rhombus Examples

❶ In rhombus $ABCD$, the measure in centimeters of $\overline{AB}$ is $3x + 2$ and $\overline{BC}$ is $2x + 9$.
Find the number of centimeters in the length of $\overline{DC}$.

Solution: A rhombus has 4 congruent sides.

Make an equation: $2x + 9 = 3x + 2$
$7 = x$
$3(7) + 2 = 23$
$AB = 23$
$\overline{AB} \cong \overline{DC}$
$CD = 23$

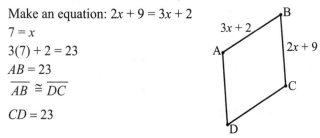

❷ The diagonals of a rhombus are 10 and 16 inches in length. Find the length of each side.

Solution: The diagonals are perpendicular and bisect each other. This forms right triangles, and we can use the Pythagorean Theorem to solve the problem. The sides of the rhombus are all congruent.

$AE = 10/2 = 5$ inches
$BE = 16/2 = 8$ inches
$\overline{AB}$ is the hypotenuse in $\triangle AEB$.
$c^2 = 5^2 + 8^2$
$c^2 = 89$
$c = \sqrt{89}$

Therefore, each side of the rhombus measures $\sqrt{89}$ inches.

Note: Since the problem did not say to round off, leave the answer in simplest radical form.

Similarity and Congruence of Polygons

Rhombus Statement Reason Proof

Given: $\angle BAC \cong \angle BCA$, $\angle DAC \cong \angle DCA$,
$\angle ABD \cong \angle ADB$, $\angle CBD \cong \angle CDB$.

Prove: $ABCD$ is a rhombus.

Statement	Reason
1. $\angle BAC \cong \angle BCA$, $\angle DAC \cong \angle DCA$, $\angle ABD \cong \angle ADB$, $\angle CBD \cong \angle CDB$.	1. Given.
2. $\angle BAC + \angle DAC \cong \angle BCA + \angle DCA$ $\angle ABD + \angle CBD \cong \angle ADB + \angle CDB$	2. Addition Axiom. (When equals are added to equals, the results are equal.)
3. $\angle BAC + \angle DAC = \angle BAD$ $\angle BCA + \angle DCA = \angle BCD$ $\angle ABD + \angle CBD = \angle ABC$ $\angle ADB + \angle CDB = \angle ADC$	3. Angle addition postulate. (The whole is equal to the sum of its parts.)
4. $\angle BAD \cong \angle BCD$ $\angle ABC \cong \angle ADC$	4. Substitution.
5. $ABCD$ is a parallelogram.	5. In a quadrilateral, if both pairs of opposite angles are congruent, it is a parallelogram.
6. In $\triangle ABD$, $\overline{AD} \cong \overline{AB}$.	6. In a triangle, if 2 angles are $\cong$, the sides opposite them are $\cong$.
7. $ABCD$ is a rhombus.	7. A parallelogram with 2 consecutive (or adjacent) sides that are congruent is a rhombus.

Show: It is a **square** (combines all the characteristics of rectangles and rhombuses).

- A parallelogram with a right angle and whose two consecutive sides are congruent is a square.

- A parallelogram whose diagonals are congruent and perpendicular to each other is a square.

- A rhombus with congruent diagonals is a square.

RIGHT TRIANGLES AND TRIGONOMETRIC RATIOS

Remember that in similar figures, corresponding angles are congruent and corresponding sides are proportional. Right triangles that are similar to each other allow us to discuss the ratios of the sides as they relate to trigonometry. Trigonometric ratios are tools that can be used to find the lengths of the sides or the measures of the acute angles in a right triangle. (At this time, the trigonometry discussion will relate to right triangles only although the trigonometric ratios can also be used in non-right triangles.)

VOCABULARY OF TRIGONOMETRY (TRIG)

Trigonometric Value: The ratio of the lengths of two sides of a triangle based on the measure of the angle involved. The trig value of an angle is unchanged without regard to where the angle is used.

Right Triangle: A triangle that contains a right angle. Indicate it on a diagram as shown Figure 1.

Acute Angles: The other two angles in a right triangle. Each is less than 90°.

Labels: The angles are labeled with upper case letters. C is often, but not always, used as the label for the 90° angle. Each side of the triangle is labeled with a lower-case letter that matches the angle across from that side. The sides can be used with their lower case label, or with the upper case segment label, but be consistent throughout a problem. $\overline{AB}$ and c are the same side.

Figure 1

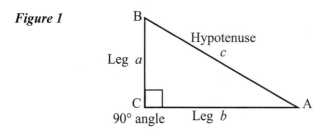

Hypotenuse: The side directly across from the right angle of a right triangle. It is the longest side of the triangle. *The hypotenuse is not considered to be "opposite" or "adjacent" to an acute angle and is simply referred to as the hypotenuse when working with trig problems.* The abbreviation used in formulas is "hyp.".

Trigonometry

Legs: The two sides of the triangle that meet to form the right angle.

Adjacent Leg or Side: The leg of the triangle that is one side of a given angle. In figure 1, $\overline{AC}$, or leg b, is adjacent to $\angle A$. Leg a or $\overline{BC}$, is adjacent to $\angle B$. The abbreviation used in formulas is "adj.".

Opposite Leg or Side: The leg of the triangle that is directly across from the acute angle. It is not part of the angle itself. The leg opposite $\angle A$ is a, or $\overline{BC}$. The side opposite $\angle B$ is b, or $\overline{AC}$. The abbreviation for opposite in formulas is "opp.".

Note: In a right triangle, it is extremely important to understand the adjacent side of one acute angle is the opposite side of the other. Leg b is adjacent to $\angle A$ but is opposite to $\angle B$.

Trigonometry Functions

For any right triangle, certain ratios are constant for the acute angles in the triangle. Using these ratios allows us to find a missing angle or a missing side when given some information about the right triangle. Abbreviated versions of the names of the ratios are used: Sine is Sin; Cosine is Cos; and Tangent is Tan.

Sin A is the ratio of the *side opposite* to the given angle (A) to the *hypotenuse* of the right triangle. $\text{Sin } A = \dfrac{\text{Opp}}{\text{Hyp}}$

Cos A is the ratio of the *side adjacent* to the given angle (A) to the *hypotenuse* of the right triangle. $\text{Cos } A = \dfrac{\text{Adj}}{\text{Hyp}}$

Tan A is the ratio of the *side opposite* to the given angle (A) to the *side adjacent* to A. $\text{Tan } A = \dfrac{\text{Opp}}{\text{Adj}}$

– Many students use "SOHCAHTOA" (**S**in = **O**pp/**H**yp, **C**os = **A**dj/**H**yp, **T**an = **O**pp/**A**dj) to help remember the trig ratios.

Using angle A as the given angle:
$$\text{Sin } A = \frac{BC}{AB} \qquad \text{Cos } A = \frac{AC}{AB} \qquad \text{Tan } A = \frac{BC}{AC}$$

These ratios are also expressed as decimal numbers that match specifically to an acute angle and its associated trig function. The ratios in that form can be found in the calculator or in a table of trig values.

Trigonometry

$$\boxed{3.4}$$

Examples

❶ Sin 25° = 0.422618261 in a calculator.
(In a table of trig values, it would show as 0.4226.)

❷ Tan 40° = 0.839099631 in a calculator.
(In a table of trig values, it would show as 0.8391.)

❸ If $a = 5$, $b = 12$, and c is unknown, write the *exact* ratios
(NO DECIMALS) for the three trig functions of angle B.

To Solve: Sketch and label a diagram. Circle the angle involved.

First find c.

$c^2 = a^2 + b^2$

$c^2 = 5^2 + 12^2$

$c^2 = 169$

$c = 13$

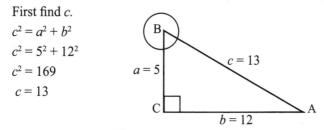

Trigonometry Functions: $Sin\ B = \dfrac{12}{13}$

$Cos\ B = \dfrac{5}{13}$

$Tan\ B = \dfrac{12}{5}$

Trigonometry

FINDING ANGLES OR TRIG RATIO VALUES

Fortunately, we don't have to do any extended arithmetic when we need to find the value of a trig ratio. The ratios are readily available in a calculator. To find the value of the trig ratio for a particular angle, use the calculator buttons labeled "sin" or "cos" or "tan". To find the value of the angle when the trig function value is already known, use the "inverses" on the calculator. These look like "sin⁻¹, cos⁻¹, or tan⁻¹ respectively.* Rounding of trig function values is often to the 10,000ᵗʰ decimal value (4ᵗʰ decimal place to the right of the decimal point) but in problem solving, it is best to use the full value in the calculator until the final step. Follow the directives of your teacher on the rounding procedure, as it can sometimes make a difference in the answers.

* Inverse trig functions can also be written using the prefix "arc".
The forms are equivalent.
Sin⁻¹ A = arcsin A
Cos⁻¹ A = arccos A
Tan⁻¹ A = arctan A

Examples

❶ $\angle A = 62°$

a) Find the value of Tan 62°. The full value that is shown in the calculator is 1.880726465. This would generally be written as Tan 62° = 1.8807.

b) Find the value of the sine of angle A. Sin 62° = 0.8829475929 which would round to 0.8829.

c) What is the Cos 62°? Cos 62° = 0.4694715648 *or* 0.4695

❷ **Cos B = 0.6427876097**

a) What is the measure of $\angle B$? The full value is given in this problem, so the result should be quite accurate. Use the cos⁻¹ function on the calculator and type in the number that is given: cos⁻¹(.6427876097) = 50°. If only the 4 digit trig value had been given, the answer might be slightly different but very close to 50°.

b) For what angle is the tangent ratio $\frac{5}{4}$?
Tan⁻¹ $\left(\frac{5}{4}\right)$ = 51.34019175 ≈ 51.34°

Geometry Made Easy – Common Core Standards Edition

RIGHT TRIANGLES, TRIGONOMETRIC RATIOS AND SIMILARITY

Right triangle similarity allows the trig ratios to be used effectively without regard to the size of the triangle. The proportionality of the corresponding sides of similar triangles combined with the congruency of corresponding angles define the trigonometric ratios for acute angles.

Example

Given: Similar right triangles *ABC* and *DEF*. The right angles are *C* and *F*. $\angle A = \angle D$; $\angle B = \angle E$. The lengths of the sides and the hypotenuse in $\triangle ABC$ and $\triangle DEF$ are in the ratio of 1:2.

Compare the trigonometric ratios of the acute angles in the two triangles.

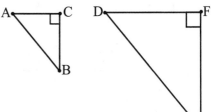

Although the triangles are not the same size, the angles are congruent. Compare the trig ratios for the two acute angles in each triangle:

$\triangle ABC$	$\triangle DEF$
$\text{Tan } A = \dfrac{BC}{AC}$	$\text{Tan } D = \dfrac{EF}{DF}$
$\text{Sin } A = \dfrac{BC}{AB}$	$\text{Sin } D = \dfrac{EF}{DE}$
$\text{Cos } A = \dfrac{AC}{AB}$	$\text{Cos } D = \dfrac{DF}{DE}$

Since the lengths of the sides are in a ratio of 1:2 we know that :
$$EF = 2BC \; ; \; DF = 2AC \; ; \; DE = 2AB$$

Using Substitution: $\text{Tan } D = \dfrac{2BC}{2AC} \; ; \text{Sin } D = \dfrac{2BC}{2AB} \; ; \; \text{Cos } D = \dfrac{2AC}{2AB}$

$$\text{Tan } D = \frac{2BC}{2AC} = \frac{BC}{AC} \quad \text{Sin } D = \frac{2BC}{2AB} = \frac{BC}{AB} \quad \text{Cos } D = \frac{2AC}{2AB} = \frac{AC}{AB}$$

Conclusion: Since the 3 trig ratios of $\angle D$ equal the 3 trig ratios of $\angle A$ respectively, using the transitive property we can say that the Tan *A* = Tan *D* ; Sin *A* = Sin *D* ; and Cos *A* = Cos *D*. So although the triangles themselves are similar triangles but different in size, it is clear that the trig ratios of the corresponding angles are equal.

The same procedure can be used to compare the trig functions of the other acute angle, *E*, and the findings will confirm the results above.

Trigonometry

Example

Given: $\triangle ABC \sim \triangle DEF \sim \triangle STU$ with right angles C, F, and U respectively.
$a = 2$ and $b = 4$

Determine the Sin B, Cos B, and Tan B. Compare your findings with Sin E, Cos E, and Tan E. What are the three trig ratios for $\angle T$?

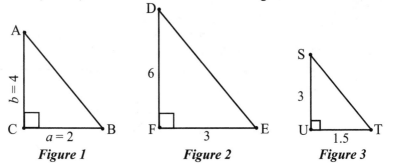

| Figure 1 | Figure 2 | Figure 3 |

Solution: Both the sine and cosine functions involve the hypotenuse, so we must find the length of AB first. Use the Pythagorean Theorem. (This is where labeling with lower-case labels can be helpful. The hypotenuse in the Pythagorean Theorem is c which in $\triangle ABC$ is $\overline{AB}$.) Then calculate the ratios. Find the hypotenuse of Figure 2.

In Figure 1:

$c^2 = a^2 + b^2$

$c^2 = (2)^2 + (4)^2$

$c^2 = 20$

$c = \sqrt{20} = 2\sqrt{5}$

Label $c = 2\sqrt{5}$

$\text{Sin } B = \dfrac{\text{Opp}}{\text{Hyp}} = \dfrac{AC}{AB} = \dfrac{4}{2\sqrt{5}} = \dfrac{2\sqrt{5}}{5} \approx .8944$

$\text{Cos } B = \dfrac{\text{Adj}}{\text{Hyp}} = \dfrac{BC}{AB} = \dfrac{2}{2\sqrt{5}} = \dfrac{\sqrt{5}}{5} \approx .4472$

$\text{Tan } B = \dfrac{\text{Opp}}{\text{Adj}} = \dfrac{AC}{BC} = \dfrac{4}{2} = 2$

(problem continued on next page)

Trigonometry

In Figure 2: Since we know the length of the hypotenuse in Figure 1, we can use either the Pythagorean Theorem again or use the similarity ratio to determine the length of the hypotenuse in Figure 2. Since there is already a radical in the problem, it may be easier to do the Pythagorean Theorem again.

Proportion

$$\frac{4}{6} = \frac{2\sqrt{5}}{x}$$

$$12\sqrt{5} = 4x$$

$$x = 3\sqrt{5}$$

P.T.

$$c^2 = a^2 + b^2$$

$$c^2 = (3)^2 + (6^2)$$

$$c^2 = 45$$

$$c = \sqrt{45} = 3\sqrt{5}$$

$$\text{Sin } E = \frac{\text{Opp}}{\text{Hyp}} = \frac{DF}{DE} = \frac{6}{3\sqrt{5}} = \frac{2\sqrt{5}}{5} \approx .8944$$

$$\text{Cos } E = \frac{\text{Adj}}{\text{Hyp}} = \frac{EF}{DE} = \frac{3}{3\sqrt{5}} = \frac{\sqrt{5}}{5} \approx .4472$$

$$\text{Tan } E = \frac{\text{Opp}}{\text{Adj}} = \frac{DF}{EF} = \frac{6}{3} = 2$$

Discussion:
The three trig functions for $\angle E$ are equal to the corresponding ratios for $\angle B$. Figure 3 is similar to Figures 1 and 2, therefore its angles are congruent to the corresponding angles in Figures 1 and 2. $\angle T$ corresponds to $\angle E$ and $\angle B$, therefore the Sin T, Cos T, and Tan T will be equal to the Sin, Cos, and Tan of both $\angle E$ and $\angle B$ respectively.

Added Thought:
Question: What is an easy way to find the length of $\overline{ST}$?

Answer: Use the ratio of Figure 1 to Figure 3 which is $\frac{4}{3}$.

$$\frac{AB}{ST} = \frac{4}{3}$$

$$\frac{2\sqrt{5}}{ST} = \frac{4}{3}$$

$$ST = \frac{3(2\sqrt{5})}{4}$$

$$ST = \frac{3\sqrt{5}}{2}$$

Trigonometry

TRIGONOMETRIC VALUES AND COMPLEMENTARY ANGLES

Complementary angles are two angles that have a sum of 90°. In a right triangle, since one angle is 90° and the sum of the three angles must be 180°, the sum of the two acute angles is always 90°. They are always complementary angles.

Considering that when working with trig ratios, the opposite and adjacent legs are identified based on the location of the angle involved, it is easy to see a relationship between the trig functions of complementary angles.

In triangle *ABC*, *C* is the right angle. The hypotenuse is labeled *c* and the legs opposite angles *A* and *B* are labeled *a* and *b* respectively. Angles *A* and *B* are complementary angles since they have a sum of 90°.

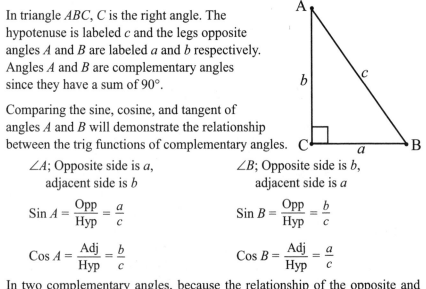

Comparing the sine, cosine, and tangent of angles *A* and *B* will demonstrate the relationship between the trig functions of complementary angles.

$\angle A$; Opposite side is *a*, adjacent side is *b*

$$\text{Sin } A = \frac{\text{Opp}}{\text{Hyp}} = \frac{a}{c}$$

$$\text{Cos } A = \frac{\text{Adj}}{\text{Hyp}} = \frac{b}{c}$$

$\angle B$; Opposite side is *b*, adjacent side is *a*

$$\text{Sin } B = \frac{\text{Opp}}{\text{Hyp}} = \frac{b}{c}$$

$$\text{Cos } B = \frac{\text{Adj}}{\text{Hyp}} = \frac{a}{c}$$

In two complementary angles, because the relationship of the opposite and adjacent sides is reversed, the trig functions of complementary angles change. The side opposite $\angle A$ is the side adjacent to $\angle B$. The sine of one angle is equal to the cosine of its complement. $\text{Sin } A = \frac{a}{c}$ and $\text{Cos } B = \frac{a}{c}$

Likewise, the cosine of one angle is equal to the sine of its complement. $\text{Cos } A = \frac{b}{c}$ and $\text{Sin } B = \frac{b}{c}$

Question: What can be said about the tangent ratios of two complementary angles? Explain.

Again, because of the reversal of the opposite vs adjacent sides of the angles, the tangent ratio changes as well. The side opposite $\angle A$ is the side adjacent to $\angle B$.

$$\text{Tan } A = \frac{\text{Opp}}{\text{Adj}} = \frac{a}{b} \text{ and Tan } B = \frac{\text{Opp}}{\text{Adj}} = \frac{b}{a}$$

The tangents of two complementary angles are reciprocals of each other.

$$\text{Tan } A = \frac{1}{\text{Tan } B} \text{ and Tan } B = \frac{1}{\text{Tan } A}$$

Trigonometry

Geometry Made Easy – Common Core Standards Edition

Examples

❶ What is the value of the sine of the angle complementary to $\angle B$ if cos $\angle B \approx .6427876097$? If two angles are complementary, the sine of one equals the cosine of the other. Since $\angle B = 50°$, its complement is 40°.

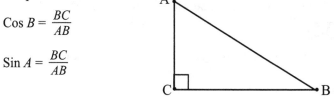

Cos $B = \dfrac{BC}{AB}$

Sin $A = \dfrac{BC}{AB}$

Cos 50° = Sin 40° = .6427876097

❷ If tan $\angle A = \dfrac{12}{7}$, what is the tan of $\angle B$ if $\angle B = (90 - A)$. Find the value of both angles to the nearest degree.

Solution: $\angle A$ and $\angle B$ are complementary. That means their tangents are reciprocals. Tan $A = \dfrac{12}{7}$, Tan $B = \dfrac{7}{12}$. To find the angles, use Tan $^{-1}\left(\dfrac{12}{7}\right)$ to find $\angle A \approx 60°$. Tan $^{-1}\left(\dfrac{7}{12}\right) \approx \angle B \approx 30°$.

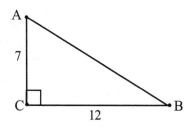

Trigonometry

SOLVING GEOMETRY PROBLEMS USING TRIGONOMETRY

To find a side of a triangle when you know one side and one acute angle:
Determine which trig function you need. Locate the angle first, then decide which two sides are involved in relation to the angle. Use the appropriate trig function, filling in the numbers that you know. Do the algebra with the formula and solve for the unknown side.

Examples

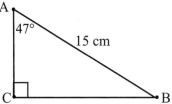

❶ Use the accompanying diagram.
If A is 47° and $AB = 15$ cm,
find the length of side BC
to the *nearest hundredth*.

$$\text{Sin } A = \frac{BC}{AB}$$

$$\text{Sin } 47 = \frac{BC}{15}$$

$(15)(\text{Sin } 47) = BC$
$BC = 10.97030552$
$BC \approx 10.97$ cm

Note: Side AB is the hypotenuse. Side BC is opposite angle A. Use Sin A. Substitute the known numbers and do the math.

The "wiggly" equal sign ($\approx$) can be used to show that the answer is approximate – it was rounded.

❷ Using the diagram from Example 1, find the length of side AC. Now we are using the adjacent side to angle A and the hypotenuse. This requires Cos A.

$$\text{Cos } A = \frac{AC}{AB}$$

$$\text{Cos } 47 = \frac{AC}{15}$$

$(15)(\text{Cos } 47) = AC$
$AC = 10.2299754$
$AC \approx 10.23$ cm

Note: Although we did find side BC in example 1 which would allow us to use Tan 47° or the Pythagorean Theorem, it is best to use the information given directly in the problem when possible. Since the answer in example 1 was rounded, it is not completely accurate which may cause a larger margin of error if it is re-used in another part of the problem. Use $\angle A = 47°$ and 15 cm given in the problem.

Trigonometry

To find an acute angle of a triangle when given the lengths of 2 sides:

Locate the angle and determine the relationship of the given sides to that angle. Set up the equation using the appropriate trig function. A decimal answer will usually be the result. This number is the actual ratio of the trig function you chose for a particular angle measure. Use the "inverse" buttons on your calculator to find the angle. They are usually located directly above the sin, cos, or tan button and require the use of the 2nd button. They look like this: $\sin^{-1}$, $\cos^{-1}$, and $\tan^{-1}$.

Example

In right triangle ABC, C is the right angle. $AC = 20$ units, $BC = 30$ units. Find the measure of angle A to the *nearest degree*. Draw a diagram. Locate the angle and the sides involved. Decide which function is needed. In this case we have the side adjacent (AC) to angle A and the side opposite (BC) angle A. Use the Tangent function.

Step:

1) Label known information.

2) Use: $\text{Tan } A = \dfrac{BC}{AC}$

3) Substitute: $\text{Tan } A = \dfrac{30}{20}$

4) Solve: $\text{Tan } A = 1.5$

5) Use inverse: $\text{Tan}^{-1}1.5 = 56.30993247$

6) Solution: $A = 56°$

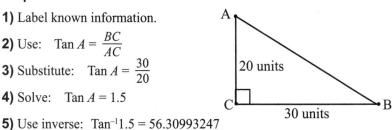

Special Right Triangles: (See also page 105)

1) In a 45°- 45°- 90° right triangle
   (Isosceles right triangle)
   - The length of the hypotenuse equals the length of either leg multiplied by $\sqrt{2}$.
   - The length of either leg = $\dfrac{1}{2}$ the hypotenuse times $\sqrt{2}$.

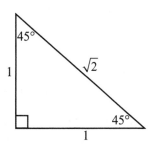

Trigonometry

2) In a 30°- 60°- 90° right triangle
- The shorter leg is = $\frac{1}{2}$ the length of the hypotenuse.
- The longer leg is = $\frac{1}{2}$ the hypotenuse times $\sqrt{3}$.
- The longer leg is equal to the shorter leg times the $\sqrt{3}$.

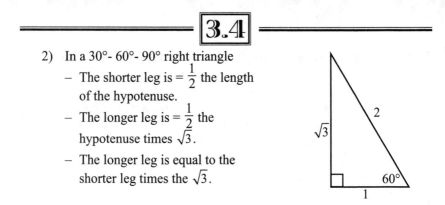

Using "given" or known information to show or find a result using right triangles. Here are some ways to connect what you know with what you need to find.

1) **Known:** Measures of two sides of a right triangle.
   Find: Measure of the third side.
   - In a right triangle, the square of the hypotenuse is equal to the sum of the squares of the two legs. Use the Pythagorean Theorem.

2) **Known:** Measure of three sides of a triangle.
   Show: The triangle is or is not a right triangle.
   - In a triangle, if the square of one side equals the sum of the squares of the other two sides, the triangle is a right triangle. [Use this theorem also to show the triangle is not a right triangle - if the sum of the squares of two sides does not equal the square of a third side, it is *not* a right triangle.]

3) **Known:** Lengths of two sides of a right triangle.
   Find: The measure of an acute angle in the triangle.
   - Use sin, cos, or tan ratios. The inverse trig button will be needed on the calculator $\sin^{-1}$, $\cos^{-1}$, or $\tan^{-1}$ to get the angle measure. Remember these are equivalent to arcsin, arccos, or arctan.

4) **Known:** The measure of an acute angle and one side of a right triangle.
   Find: Another side of the triangle.
   - Use sin, cos, or tan ratios.

5) **Known:** One acute angle in a right triangle.
   Find: The other acute angle.
   - The acute angles in a right triangle are complementary. Remember all three angles must add up to 180°. Since one angle is 90°, the sum of the other two angles is 90°.

Trigonometry

Special Right Triangles: Right triangles that have angles of 30°, 60°, and 90° or 45°, 45°, and 90° contain angles that are very commonly found in our work with trigonometry. If these two diagrams are memorized, it is easy to find the exact trigonometry values for 30°, 45°, and 60° angles (or angles of $\dfrac{\pi}{6}$, $\dfrac{\pi}{4}$, and $\dfrac{\pi}{3}$ radians).

Examples

❶ Right triangle with angles of 30°, 60°, and 90° or $\dfrac{\pi}{6}$, $\dfrac{\pi}{3}$, and $\dfrac{\pi}{2}$ radians.

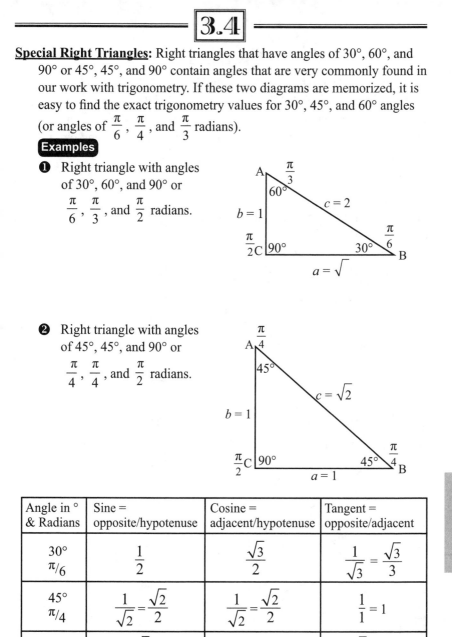

❷ Right triangle with angles of 45°, 45°, and 90° or $\dfrac{\pi}{4}$, $\dfrac{\pi}{4}$, and $\dfrac{\pi}{2}$ radians.

Angle in ° & Radians	Sine = opposite/hypotenuse	Cosine = adjacent/hypotenuse	Tangent = opposite/adjacent
30° $\pi/6$	$\dfrac{1}{2}$	$\dfrac{\sqrt{3}}{2}$	$\dfrac{1}{\sqrt{3}} = \dfrac{\sqrt{3}}{3}$
45° $\pi/4$	$\dfrac{1}{\sqrt{2}} = \dfrac{\sqrt{2}}{2}$	$\dfrac{1}{\sqrt{2}} = \dfrac{\sqrt{2}}{2}$	$\dfrac{1}{1} = 1$
60° $\pi/3$	$\dfrac{\sqrt{3}}{2}$	$\dfrac{1}{2}$	$\dfrac{\sqrt{3}}{1} = \sqrt{3}$

The trig ratios for the special right triangles are common in many problems. It is very important to understand their origins, and it is helpful to be able to recall them from memory.

Trigonometry

TRIG HAND

This is a memory trick to help remember the special angle trig function values. *It does not provide understanding of the topic.*

The thumb represents 90° since it makes a "right" angle, sort of, with the hand. The index finger is 60°, long finger is 45°, ring finger is 30°, and pinky is 0°. The hand itself is 2. Fold the finger that represents the angle into the palm of the hand.

BACK of Left Hand – SINE: Divide the square root of the number of fingers above the folded finger by 2.

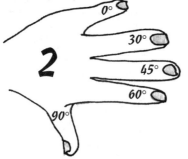

Pinky: $\dfrac{\sqrt{0}}{2} = 0 = \sin 0°$

Ring: $\dfrac{\sqrt{1}}{2} = \dfrac{1}{2} = \sin 30°$

Long: $\dfrac{\sqrt{2}}{2} = \sin 45°$

Index: $\dfrac{\sqrt{3}}{2} = \sin 60°$

Thumb: $\dfrac{\sqrt{4}}{2} = 1,\ \sin 90° = 1$

PALM of Left Hand – COSINE: Divide the square root of the number of fingers above the folded finger by 2.

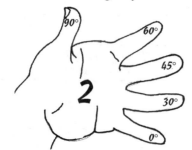

Thumb: $\dfrac{\sqrt{0}}{2} = 0 = \cos 90°$

Index: $\dfrac{\sqrt{1}}{2} = \dfrac{1}{2} = \cos 60°$

Long: $\dfrac{\sqrt{2}}{2} = \cos 45°$

Ring: $\dfrac{\sqrt{3}}{2} = \cos 30°$

Pinky: $\dfrac{\sqrt{4}}{2} = 1 = 1 = \cos 0°$

Example

If the ring finger is folded into the palm there will be 3 fingers above the ring finger.

The cosine of 30° is $\dfrac{\sqrt{3}}{2}$.

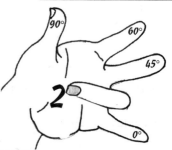

Geometry Made Easy – Common Core Standards Edition

Unit 4

EXTENDING TO THREE DIMENSIONS

- Explain volume formulas and use them to solve problems.

- Visualize relationships between two-dimensional and three-dimensional objects.

- Apply geometric concepts in modeling situations.

Cylinder: The volume of a prism is, in general, the area of the base multiplied by the height of the prism. In the case of a cylinder, the base is a circle. Since the area of a circle is known to be $A = \pi r^2$, as is demonstrated on page 142, this area formula needs to be multiplied by the perpendicular distance between the bases of the cylinder, called the altitude or the height.

Example What is the volume, in cubic inches, of a can of coffee that has a diameter of 10 inches and is 8 inches tall? Remember to use the radius of 5 in the circle area formula.

$V = Bh$

$V = \pi r^2 h$

$V = \pi (5)^2 8$

$V = 200\pi$

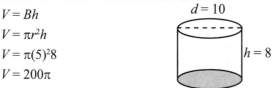

$d = 10$

$h = 8$

Rectangular Solid: Since the faces of a rectangular solid are all rectangles, choosing one face as the base, finding its area using $A = lw$, then multiplying by the length of the side perpendicular to it will produce the area of the rectangular prism. $A = lwh$.

Example

$V = lwh$

$V = (5)(6)(8)$

$V = 240$ cubic units.

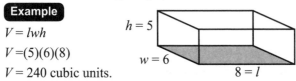

$h = 5$

$w = 6$

$8 = l$

Cavalieri's Principle: This is another way to consider the volume of cylinders, prisms, and other 3 dimensional figures.

If two figures have equal altitudes and equal areas for each parallel plane that can be cut through them, their volumes are equal. Think of a stack of coins that is in a straight pile. Each coin represents a parallel plane. Then tilt the stack to one side. Each coin in the tilted pile has the same area as it counterpart in the straight pile, and both stacks have the same altitude. Cavalieri's Principle tells us that the volume for either stack of coins is the product of the base area and the altitude and that the two stacks have equal volume.

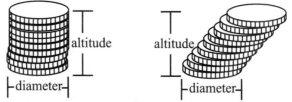

altitude

altitude

diameter

diameter

Geometric Measurement

Cones and Pyramids: These are related to cylinders and rectangular solids. To informally demonstrate the relationship between cones and cylinders or pyramids and rectangular solids, it is best to do a physical demonstration of pouring rice (or water) from a cone into a cylinder with the same base and height. It will take three cones full of the rice to fill the cylinder. Using a pyramid with a rectangular base and doing the same procedure to fill a rectangular solid with the same base and height requires three pyramids filled with rice to fill the rectangular solid. The formulas for the volumes of the cone and pyramid are each $\frac{1}{3}$ of their associated solid.

Cone: $V = \frac{1}{3} Bh$ where $B = \pi r^2$ **Pyramid:** $V = \frac{1}{3} Bh$ where $B = lw$

The formula for the volume of the pyramid can be fairly easily demonstrated further using a cube. The cube can be "sliced" into three congruent pyramids. They have congruent bases and altitudes. Since they are congruent, they have equal volumes. That makes each one equal to $\frac{1}{3}$ the volume of the cube.

Examples **Cones and Pyramids**

❶ **Cone:** The radius of the base of a cone is 9. The height or altitude of the cone is 10. What is the volume of the cone?

$A_{circular\ base} = \pi r^2$

$A = (9)(9)\pi$

$V = \frac{1}{3} Bh$

$V_{cone} = \frac{1}{3}(81\pi)(10)$

$V = 270\pi$

$V_{cone} = 848.23$ cubic units

❷ **Pyramid:**

$A_{Base} = s^2$

$A_{Base} = (10)^2 = 100$

To find h, use the Pythagorean Theorem.

Note: 5-12-13 is a Pythagorean Triple

$13^2 = s^2 + h^2$

$169 = 25 + h^2$

$\sqrt{144} = h$

$h = 12$

$V_{pyramid} = \frac{1}{3} Bh *$

$V = \frac{1}{3}(100)(12)$

$V_{pyramid} = 400$ cubic units

* Since B equals s^2, this formula could be written $V = s^2 h$

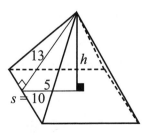

$$\boxed{4.1}$$

SPHERES

Sphere: A sphere is developed by rotating a circle about any of its diameters. The formula for the volume of a sphere is $V = \dfrac{4}{3}\pi r^3$.

Examples

❶ What is the volume, to the *nearest tenth*, of a sphere with a diameter of 9 inches?

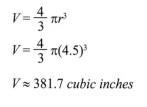

$V = \dfrac{4}{3}\pi r^3$

$V = \dfrac{4}{3}\pi(4.5)^3$

$V \approx 381.7$ *cubic inches*

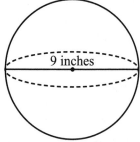

9 inches

❷ Andrew has a box shaped like a cube that contains sand. He is going to transfer the sand to spherical containers for a science project. The box is 14 inches on each side. The containers are spheres with the diameter equal to 8 inches. How many spheres will he be able to fill completely?

Solution: Find the volume of the cube and the volume one of the spherical containers will hold. Divide the total amount of sand by the volume for each sphere.

Cube:

$V = s^3$

$V = (14)^3$

$V = 2744$ *cubic inches of sand*

Sphere:

$V = \dfrac{4}{3}\pi r^3$

$V = \dfrac{4}{3}\pi(4)^3$

$V \approx 268.08$ *cubic inches*

Number of spherical containers: $\dfrac{V_{cube}}{V_{sphere}} = \dfrac{2744}{268.08} = 10.24$

Conclusion: Andrew will be able to fill 10 spherical containers completely.

Geometric Measurement

Geometry Made Easy – Common Core Standards Edition

3-DIMENSIONAL FIGURES AND THEIR PROPERTIES

Solid: A closed figure, usually including its interior space.

Polyhedron: A closed figure with faces that are planar (flat). The faces meet on line segments called **edges** and the edges meet at points called **vertices**.

Cross Section: A cross section is a 2-dimensional figure that is created when a plane is passed through a polyhedron. The cross section can be many different shapes, depending on the polyhedron involved. Cross sections of solids related to circles are also created when a plane passes through a cylinder or sphere.

Parallelepiped: A polyhedron that has parallel bases and all the lateral faces are parallelograms. The volume of a parallelepiped equals the product of the area of the base (B) and the altitude (h) which is the perpendicular distance between the two parallel bases.

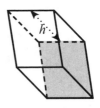

Some Parallelepipeds have specific characteristics that classify them as prisms.

Prism: A polyhedron with parallel congruent bases which are both polygons and lateral faces that are all parallelograms. The lateral edges of a prism are congruent and parallel. The height of a prism is the perpendicular distance between the parallel bases.

Cube: A prism whose faces are all congruent squares. (Ex: A die)

Cross sections of a cube:

 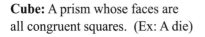

triangle   square   rectangle   trapezoid   pentagon   hexagon

Geometric Measurement

Rectangular or Right Prism: All its faces are rectangles. (Ex: A shoe box)

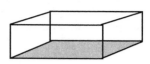

Cross sections of a rectangular prism, or solid, are the same as those for a cube. In order to obtain a square cross section, the plane must be angled.

Triangular Prism: The 2 parallel bases are triangles, the other sides are rectangles. (Ex: a pup tent)

Cross Sections of a triangular prism:

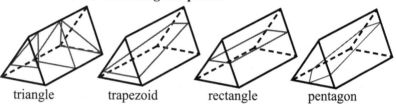

triangle   trapezoid   rectangle   pentagon

Pyramid: A polyhedron with a polygon for its base. The lateral sides are triangles and they meet at a point called a vertex or apex. The volume of a pyramid is 1/3 the product of the area of the base and the altitude which is the perpendicular distance from the apex to the base.

Regular Pyramid: The base is a regular polygon (congruent sides, congruent angles). Its lateral sides or faces are congruent isosceles triangles and its lateral edges are congruent. When the altitude, *h*, is dropped from the apex (or vertex), it meets the base at its center. The *slant height*, *s*, of a regular pyramid is the altitude of one of the congruent isosceles triangular sides. The base can be any type of regular polygon.

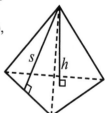

Cross sections of a pyramid:

triangle   trapezoid

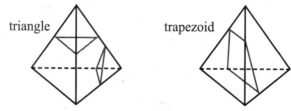

Geometric Measurement

Cylinders and Cones have circular bases and curved lateral sides.

Cylinder: This figure contains two congruent circular bases that are parallel. The lateral surface is a curve that connects the circumference of one base to the circumference of the other. (Although other types of cylinders do exist, this is a right cylinder.) Think of a can of soup.

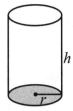

Cross sections of a cylinder:

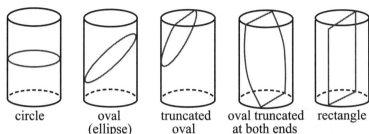

| circle | oval (ellipse) | truncated oval | oval truncated at both ends | rectangle |

Cone: Has a circular base and a curved surface for its "sides" which meets in a single vertex opposite the base. We will work with a right cone – one in which the altitude, h, of the cone meets the circular base at its center. With a cone we again sometimes need the *slant height*, s, which is the length of a segment drawn from the vertex to the perimeter of the base.

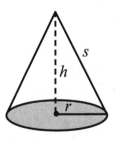

Geometric Measurement

Cross sections of a cone:

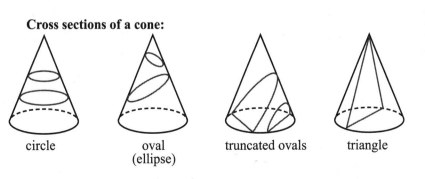

| circle | oval (ellipse) | truncated ovals | triangle |

Sphere: A three dimensional curved figure made up of all the points equidistant from the center. A ball is a sphere.

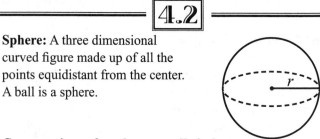

Cross sections of a sphere are all circles:

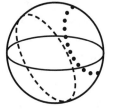

Great Circle: If a plane intersects a sphere and passes through its center, the intersection of the sphere and the plane is a great circle. A great circle is the largest circle that can be drawn on the sphere. It has the same radius as the sphere.

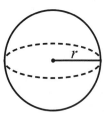

Note: Great circles are used in the navigation industry. The shortest distance between any 2 points on a sphere (the Earth) is along the circumference of the great circle containing them. Airlines and ships often use great circle navigation.

- If two planes intersect a sphere at equal distances from the center, two congruent circles are formed. They have equal radii.

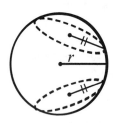

Geometric Measurement

CHANGING A 2-DIMENSIONAL FIGURE TO A 3-DIMENSIONAL FIGURE

Rotation of a 2-dimensional object can create a 3-dimensional object. A rotation of a triangle can create a cone. Rotating a rectangle or square can create a cylinder.

Examples

❶ Rectangle *ABCD* is rotated 360° about its length, *AB*. This rotation produces a cylinder with a radius equal to *AD* and height equal to *AB*.

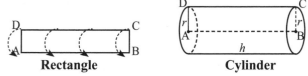

Rectangle   **Cylinder**

❷ An isosceles triangle rotated about its altitude forms a cone with altitude equal to the altitude of the triangle. The radius of the circular base is equal to one-half the base of the triangle, and the slant height is equal to one of the two congruent sides.

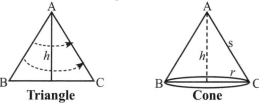

Triangle   **Cone**

❸ A circle rotated about its diameter forms a sphere with the same radius as the circle. The center of the circle becomes the center of the sphere.

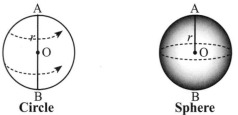

Circle   **Sphere**

❹ Combinations of 2-dimensional figures can be rotated to form 3-dimensional figures as well. A rectangle with an attached triangle can be rotated to form a shape that resembles a sharpened round pencil. The line of rotation is ℓ in this diagram.

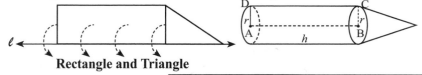

Rectangle and Triangle

Geometric Measurement

CONNECTING
ALGEBRA AND
GEOMETRY THROUGH
COORDINATES

Coordinate or Analytic Geometry is the work involved in associating a graph with algebraic processes. Formulas and equations are used along with logical reasoning skills to determine the classification of a figure, compare figures, find the dimensions of polygons or segments, and prove the congruency or similarity of figures drawn on a graph.

- Use coordinates to prove simple geometric theorems algebraically.

    - Parallel and perpendicular lines
    - Prove geometric figures algebraically
    - Partitioning a segment
    - Perimeter and area

Each point on a plane corresponds to one specific location named by the appropriate ordered pair of numbers in the form (x, y). A rectangular coordinate system is usually shown on a "grid" that represents the plane. The plane is also called a Cartesian Plane. The x value of an ordered pair is called the **abscissa**. The y value is called the **ordinate**. The correspondence of each point in a plane to an ordered pair in a coordinate system allows us to locate points on the plane. Using a "grid" or coordinate graph, we can use algebraic laws and processes along with the geometry "reasons" shown previously to solve problems. The horizontal axis of the graph is the x-axis and the vertical axis is the y-axis. The coordinate graph is understood to have a scale of 1:1 unless labeled otherwise. The intersection of the two axes at $(0, 0)$ is called the **origin**. The x-axis and y-axis are accepted as being perpendicular to each other.

Y-Intercept: The y value at the point where the graphed line crosses the y-axis. It can be found by substituting "0" for x in the given equation and finding the value for y. (Remember that everywhere on the y-axis, $x = 0$.) The y-intercept point is often used as a starting point when graphing equations from the slope-intercept form of an equation.

Example Find the y-intercept: $2y = 4x + 12$

Substitute: $2y = 4(0) + 12$

$2y = 12$

The y-intercept is 6. $y = 6$

This means the graph crosses the y-axis at $(0, 6)$.

_or_ Use the slope-intercept form of the equation, $y = mx + b$.

Isolate y, and set the equation up in the form $y = mx + b$;

b is the y-intercept. Since $b = 6$, the point the line crosses the y-axis is $(0, 6)$.

$$\frac{2y}{2} = \frac{4x + 12}{2}$$
$$y = 2x + 6$$
$$b = 6$$

Expressing Geometric Properties with Equations

X-Intercept: The x value of the point where a graphed line crosses the x-axis. The y value for the x-intercept is always 0. The x-intercept is not as commonly used but for some types of problems it is essential. Ex: A graph that represents an equation for the tossing of a ball in the air has an x value for time from toss, y value is the height of the ball. The graph begins at the y-intercept when it is tossed and crosses the x-axis when the ball hits the ground. The x value represents the time after it is thrown that the ball hits the ground.

Example Find the x-intercept: $\qquad 5y = 15x - 75$

Substitute 0 for y and solve for x. $\quad 5(0) = 15x - 75$

$$75 = 15x$$

The x-intercept is 5. $\qquad\qquad\qquad x = 5$

The point of intersection of the graph with the x-axis is $(5, 0)$.

Slope: Slope tells the "steepness and direction" of the slant of a line on a graph. It is the ratio of the vertical change in the line to the horizontal change in the line. (In science classes, slope is often described as "rise over run.") Slope is written as a fraction. Although m is used to represent the slope in formulas, it is necessary to write the word "slope" when the final answer to the formula is found or when slope is used in other ways. Use the slope to find points on the line. The top (numerator) of the fraction tells the vertical movement along the line from one point to another. The bottom (denominator) of the fraction shows the horizontal movement along the line between the points.

Example Negative slope $\qquad\qquad$ Positive Slope

$$-\frac{3\downarrow}{4\rightarrow} \qquad\qquad\qquad \frac{3\uparrow}{4\rightarrow}$$

4 units to the right, $\qquad\qquad$ 4 units to the right,
3 units down. $\qquad\qquad\qquad\quad$ 3 units up.

Hint: The "upper" part of the slope fraction tells the "up or down" movement on the graph. Write arrows on your slope fraction to help you remember.

Expressing Geometric Properties with Equations

Slope Formula: $\boxed{m = \dfrac{y_2 - y_1}{x_2 - x_1}}$ also as shown: $\boxed{m = \dfrac{\Delta y}{\Delta x}}$

Slope Formula: Where (x_2, y_2) is one point on the line and (x_1, y_1) is another point on the line.

Example Find the slope of the line through (4, 7) and (–2, –3).

$m = \dfrac{-3 - 7}{-2 - 4} = \dfrac{-10}{-6} = \dfrac{5}{3}$ (3 units to the right and up 5 units when used for graphing).

Slope and Equations: Slope can also be found as the coefficient of x when a linear equation is written in the form $y = mx + b$. m represents the slope.

Example $y = 5x - 4$. Slope is 5 or $\dfrac{5}{1}$.

(1 unit right and 5 units up when graphing.)

Point-Slope Form of an Equation: To write the equation of a line, if you know the slope and a point on the line, use this formula: $y - b = m(x - a)$ where m is the slope of the line and (a, b) is a point on the line.

Example Write the equation of a line going through the point (–4, 5) and having a slope of $\dfrac{3}{4}$.

Steps

1) Substitute. $\qquad y - 5 = \dfrac{3}{4}(x - (-4))$

2) Multiply. $\qquad y - 5 = \dfrac{3}{4}(x + 4)$

$\qquad\qquad\qquad y - 5 = \dfrac{3}{4}x + 3$

3) Isolate y. $\qquad \underline{ + 5 \qquad\quad + 5}$

4) Answer. $\qquad\qquad y = \dfrac{3}{4}x + 8$

Expressing Geometric Properties with Equations

Special Slopes: MEMORIZING THEM is best.

Vertical Lines: Slope is undefined.

> **Example** $x = 3$ is a vertical line crossing the x-axis at 3. It is parallel to the y-axis. Two points on the line are $(3, 0)$ and $(3, 2)$. Using $m = \dfrac{\Delta y}{\Delta x}$, causes the denominator to be zero. The slope is undefined since division by zero is undefined.

Horizontal Lines: Slope $= 0$.

> **Example** $y = -4$ is a horizontal line crossing the y-axis at -4. It is parallel to the x-axis. $y = -4$ goes through $(0, -4)$ and $(5, -4)$. $m = \dfrac{\Delta y}{\Delta x}$; so the numerator of the slope fraction is zero. Therefore, the slope $= 0$.

Collinear points: Two or more points that are on the same line. If the slope is the same between the first and second points as it is between the second and third (or first and third), then the points are on the same line. Use the slope formula to test.

Examples

❶ Are the points $A(3, 5)$, $B(6, 8)$, and $C(-4, -8)$ collinear?

Solution: First, find the slope between points A and B. Then use the same formula for points B and C.

Slope of points A and B	Slope between points B and C
$m = \dfrac{8 - 5}{6 - 3} = \dfrac{3}{3} = 1$	$m = \dfrac{-8 - 8}{-4 - 6} = \dfrac{-16}{-10} = \dfrac{8}{5}$

Since the slope between the first two points is 1, and the slope between the second and third points is $\dfrac{8}{5}$, thus not the same, these three points are *not* on the same line.

❷ Are the points $A(-5, -5)$, $B(0, 5)$, and $C(1, 7)$ collinear?

Slope of points A and B	Slope between points B and C
$m_{\overline{AB}} = \dfrac{-5 - 5}{-5 - 0} = \dfrac{-10}{-5} = 2$	$m_{\overline{BC}} = \dfrac{7 - 5}{1 - 0} = \dfrac{2}{1} = 2$

The slope between each pair of points is equal. The points *are* collinear.

PARALLEL AND PERPENDICULAR LINES

Mathematically correct thinking processes must be used in coordinate geometry to analyze relationships between lines. Below are some that are used with the slope formula, or the equation form of the line ($y = mx + b$), to prove whether or not lines are parallel ($\parallel$) or perpendicular ($\perp$).

- If two lines have the <u>same slope</u>, the lines are parallel.

  Example Are $y = 3x - 4$ and $y = 3x + 7$ parallel lines? They both have a slope of 3, so yes, they are parallel lines.

- If two lines are <u>parallel</u>, they have the same slope.

  Example If a line has a slope of -3, write the equation of another line that is parallel to it: $y = -3x + 1$. Both lines have a slope $= -3$, so they are parallel.

- If two lines have slopes that are <u>negative reciprocals</u> of each other, the lines are perpendicular.

  Example Are these two lines perpendicular?
  $y = 3x - 4$ and $y = -\dfrac{1}{3}x + 7$. Since the slopes are 3 and $-\dfrac{1}{3}$ which are negative reciprocals of each other, the lines are perpendicular.

- If two lines are <u>perpendicular</u>, their slopes are negative reciprocals of each other.

  Example Name two points that would be on a line that is perpendicular to the line $y = -2x$. To solve this, we must find two points that will work in the slope formula to make a slope of $\dfrac{1}{2}$. One answer might be (6, 3) and (12, 6).

Note: When two numbers that are negative reciprocals of each other are multiplied together, their product is -1.

Expressing Geometric Properties with Equations

Writing the Equation for parallel or perpendicular lines

- Write the equation of a line perpendicular to a given line through a given point on the line. Find the slope of the original line. The slope of the line perpendicular to it is the negative reciprocal of that number. The x and y values of the point are used to find the y-intercept of the new line.

Example Write the equation of a line perpendicular to the line $y = 2x + 3$ that goes through the point $(6, 9)$.

Solution: The slope of the given line is 2. Therefore the slope of a line perpendicular to it will be $(-1/2)$. Use $(-1/2)$ in the slope intercept form of an equation and substitute the values $(6, 9)$ in it to find the value of b. Rewrite the new equation using x and y.

$$9 = -\frac{1}{2}(6) + b$$
$$9 = -3 + b$$
$$b = 12$$

Answer: $y = -\dfrac{1}{2}x + 12$

- Write the equation of a line that is the perpendicular bisector of a line segment. Same procedure as the above example, but we must find the midpoint of the segment first.

Example Write the equation of the line that is the perpendicular bisector of the line segment joining the points $(-2, -4)$ and $(4, -6)$.

Solution: Find the slope of the given line segment. Find the midpoint of the given segment. Use the same method we used in the above example to write the equation.

Slope of given line	Midpoint of given segment (see page 126)
$m = \dfrac{y_2 - y_1}{x_2 - x_1}$	$M = \left(\dfrac{x_1 + x_2}{2}, \dfrac{y_1 + y_2}{2}\right)$
$m = \dfrac{-6 - (-4)}{4 - (-2)}$	$M = \left(\dfrac{-2 + 4}{2}, \dfrac{-4 + (-6)}{2}\right)$
$m = -\dfrac{1}{3}$	$M = (1, -5)$

The slope of the perpendicular bisector will be 3. It will go through the midpoint, $(1, -5)$.

$$-5 = 3(1) + b$$
$$b = -8$$

Answer: $y = 3x - 8$

Expressing Geometric Properties with Equations

- Write the equation of a line parallel to another line through a specific point. This is similar to the first example on the previous page, although the new equation must have the same slope as the given line.

  Example Write the equation of a line parallel to the line $y = 3x - 5$ that goes through the point (–3, 4).

  Slope of the given line is 3. The slope of the new line is also 3.
  Substitute (–3, 4) to find b.

  $$4 = 3(-3) + b$$
  $$b = 13$$

  Answer: $y = 3x + 13$

  When developing equations of lines parallel or perpendicular to another line, be careful to pay attention to the slope needed. Sketch the problem on graph paper or check it in your graphing calculator to see that the equations do actually produce lines that are parallel or perpendicular to each other.

DISTANCE AND MIDPOINT

Length of a Line Segment: Finding the length of a line segment is necessary to prove whether line segments are congruent or not. The length is also needed to find the area of geometric figures on a graph. If the line segment is not horizontal or vertical, the distance formula is needed. (See page 126)

A horizontal segment is *parallel to the x-axis* and has a length equal to the absolute value of the difference of the x values of the end points of the segments. Length = $|x_2 - x_1|$.

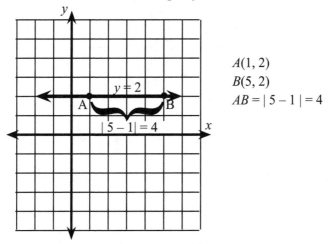

$A(1, 2)$
$B(5, 2)$
$AB = |5 - 1| = 4$

A vertical segment is *parallel to the y-axis* and has a length equal to the absolute value of the difference of the y values of the endpoints of the segments. Length = $|y_2 - y_1|$.

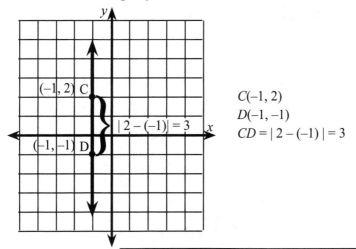

$C(-1, 2)$
$D(-1, -1)$
$CD = |2 - (-1)| = 3$

Expressing Geometric Properties with Equations

Distance Formula: This formula is used to find the length of a line segment that is *not parallel* to either axis. Using the (x, y) values for both endpoints, the formula is $\boxed{d = \sqrt{(x_2 - x_1)^2 + (y_2 - y_1)^2}}$.

Example Find the length of the line segment joining the points $(4, 6)$ and $(1, 2)$. Consider $(4, 6)$ to be point 1. $(1, 2)$ is then point 2.

$$d = \sqrt{(1 - 4)^2 + (2 - 6)^2} = \sqrt{(-3)^2 + (-4)^2} = \sqrt{9 + 16} = \sqrt{25} = 5$$

Note: The distance formula is used to find areas and perimeters of polygons on a coordinate plane.

Midpoint of a Line Segment: The midpoint is used to show lines segments are congruent, to show that a line segment is bisected by another line, and to assist in proving many kinds of problems. The coordinates of the midpoint of a line segment are found by using the following formula with the x and y values of the endpoints of the segment:

$$\boxed{\text{Midpoint Formula: } (x, y)_{\text{midpoint}} = \left(\frac{x_2 + x_1}{2}, \frac{y_2 + y_1}{2} \right)}$$

Example Find the coordinates of the midpoint of the segment joining $(-2, 4)$ and $(8, 6)$.

$$x = \frac{-2 + 8}{2} \Rightarrow \frac{6}{2} \Rightarrow 3 \quad \text{and} \quad y = \frac{4 + 6}{2} \Rightarrow \frac{10}{2} \Rightarrow 5$$

The coordinates of the midpoint are $(3, 5)$.

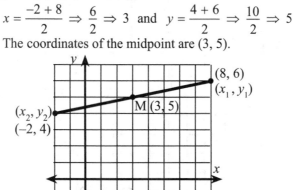

Geometry Made Easy – Common Core Standards Edition

Expressing Geometric Properties with Equations

5.2

COORDINATE OR ANALYTIC PROOF EXAMPLE

Use the slope, distance, and midpoint formulas to prove the relationships between the sides of a given figure to make a conclusion. Several examples follow, but remember that there are usually several ways to solve each one. *Your work must be done with your textbook and your own teacher's instructions in mind.*

Note: Good organization and clearly written conclusions are necessary.

Examples

❶ Given the quadrilateral *ABCD* with its vertices at *A*(3, –5), *B*(5, 1), *C*(1, 4) and *D*(–5, 1). Determine (prove) whether or not *ABCD* is a trapezoid, and if $\overline{CD}$ is congruent to $\overline{AB}$.

To solve, draw a diagram on graph paper and make a plan for your work: First prove that the diagram is a trapezoid by showing that two sides are parallel using the slope formula. $\overline{CB}$ is parallel to $\overline{DA}$. Use this distance formula to find the length of $\overline{CD}$ and then the length of $\overline{AB}$ and compare.

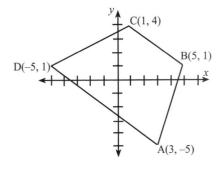

Slope formula: $m = \dfrac{y_2 - y_1}{x_2 - x_1}$ **Distance Formula:** $d = = \sqrt{(x_2 - x_1)^2 + (y_2 - y}$

$$m_{CB} = \frac{4-1}{1-5} = \frac{3}{-4} = -\frac{3}{4}$$

$$d_{CD} = \sqrt{\left(1 - (-5)\right)^2 + (4-1)^2}$$

$$m_{DA} = \frac{1-(-5)}{-5-3} = \frac{6}{-8} = -\frac{3}{4}$$

$$d_{CD} = \sqrt{(1+5)^2 + (3)^2} = \sqrt{36+9} = \sqrt{45}$$

$$m_{CD} = \frac{4-1}{1-(-5)} = \frac{3}{6} = \frac{1}{2}$$

$$d_{AB} = \sqrt{(5-3)^2 + \left(1 - (-5)\right)^2}$$

$$m_{AB} = \frac{-5-1}{3-5} = \frac{-6}{-2} = 3$$

$$d_{AB} = \sqrt{(2)^2 + (1+5)^2} = \sqrt{4+36} = \sqrt{40}$$

Conclusion: $\overline{CB} \parallel \overline{DA}$ because they have equal slopes. Quadrilateral *ABCD* is a trapezoid because it has one pair of parallel sides. $\overline{CD}$ has a length of $\sqrt{45}$ and $\overline{AB}$ has a length of $\sqrt{40}$ showing that the two non-parallel sides are not equal.

Expressing Geometric Properties with Equations

❷ **Given:** Quadrilateral *ABCD* with vertices at *A*(–4, 2), *B*(2, 6), *C*(6, 0) and *D*(0, –4) and diagonals $\overline{AC}$ and $\overline{BD}$ which intersect at *E*.

Prove: *ABCD* is a rhombus.

Plan: 1) Prove opposite sides are parallel to show it is a parallelogram. Use slope formula.

2) Prove diagonals are perpendicular - slopes are negative reciprocals of each other; or prove two adjacent sides are equal in length using the distance formula.

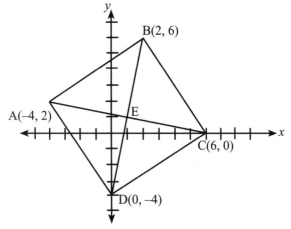

Slope Formula: $m = \dfrac{y_2 - y_1}{x_2 - x_1}$

Slope $AB = \dfrac{6-2}{2-(-4)} \Rightarrow \dfrac{4}{6} \Rightarrow \dfrac{2}{3}$ Slope $DC = \dfrac{-4-0}{0-6} \Rightarrow \dfrac{-4}{-6} \Rightarrow \dfrac{2}{3}$

Slope $AD = \dfrac{-4-2}{0-(-4)} \Rightarrow \dfrac{-6}{4} \Rightarrow -\dfrac{3}{2}$ Slope $BC = \dfrac{6-0}{2-6} \Rightarrow \dfrac{6}{-4} \Rightarrow -\dfrac{3}{2}$

Slope $\overline{AB}$ = Slope $\overline{DC}$ so $\overline{AB} \parallel \overline{DC}$, and slope $\overline{AD}$ = slope $\overline{BC}$, showing that $\overline{AD} \parallel \overline{BC}$. *ABCD* is a parallelogram because opposite sides are parallel.

Slope $BD = \dfrac{-4-6}{0-2} \Rightarrow \dfrac{-10}{-2} \Rightarrow 5$ Slope $AC = \dfrac{2-0}{-4-6} \Rightarrow \dfrac{2}{-10} \Rightarrow -\dfrac{1}{5}$

Conclusion: 5 and $-\dfrac{1}{5}$ are negative reciprocals of each other. Therefore, $\overline{BD} \perp \overline{AC}$. *ABCD* is a rhombus because it is a parallelogram with perpendicular diagonals.

PARTITIONING A SEGMENT IN A GIVEN RATIO

Several methods can be used to partition a segment into a given ratio. Use the midpoint and analyze step by step, use a formula for partitioning, or use construction.

MIDPOINT

Finding the midpoint of a segment divides the segment into two equal parts. The segments have a ratio of 1:1. Each is 1 of 2 parts that make up the whole segment.

To find the points on a segment that divides the segment into a specific ratio, sometimes using the midpoint formula multiple times works fine, other times a different method is needed.

Examples **Using The Midpoint Formula**

❶ Partition (divide) $\overline{AB}$ into 2 parts in a ratio of 1:3. The endpoints are $A(-3, 1)$ and $B(5, 3)$. This means that the first part of the ratio begins with the segment that starts with A and the second part ends with the segment that ends at B.

Analysis: Start at the first letter of segment that is named. Since the ratio of 1:3 is "1 part to 3 parts of the whole", we can think of this as making the segment into 4 parts and then using 1 part as the 1 and the other 3 as the 3 part of the ratio.

Procedure: Find the midpoint of segment $\overline{AB}$ and label it C. Find the midpoint of $\overline{AC}$ and label it D. Point D is the point that partitions $\overline{AB}$ into 2 parts with a 1:3 ratio. Make a sketch to clarify your thinking.

Midpoint:

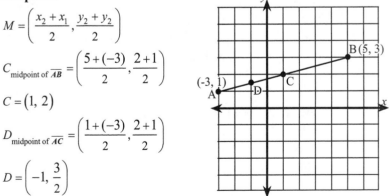

$$M = \left(\frac{x_2 + x_1}{2}, \frac{y_2 + y_1}{2} \right)$$

$$C_{\text{midpoint of } \overline{AB}} = \left(\frac{5 + (-3)}{2}, \frac{2 + 1}{2} \right)$$

$$C = (1, 2)$$

$$D_{\text{midpoint of } \overline{AC}} = \left(\frac{1 + (-3)}{2}, \frac{2 + 1}{2} \right)$$

$$D = \left(-1, \frac{3}{2} \right)$$

Conclusion: The point $D(-1, \frac{3}{2})$ partitions $\overline{AB}$ into 2 segments in the ratio 1:3. $\overline{AD} : \overline{DB} = 1:3$

Expressing Geometric Properties with Equations

STEP BY STEP AND FORMULA PARTITIONING

When dividing a segment into more complicated ratios than Example 1, a different procedure is needed. (It can, of course, be used on the simple ones as well.) In this procedure we are finding the values need to be added to the (x, y) values of the starting point to create a point that will partition the segment into the ratio given. Example 2 shows the step by step work needed, and Example 3 shows the formula that can be used to accomplish the partitioning.

❷ **Example of a Step-by-Step Procedure**

Determine the coordinates of a point, C, that will divide $\overline{AB}$ into segments in the ratio of 3:2.

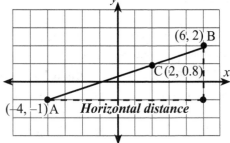

Analysis: In this problem we need a total of 5 parts that we can group together in a 3 parts: 2 parts relationship. Start with Point A to develop the 3 part segment. The 2 part segment will end with B. How much do we need to add to x_1 and y_1 to find C ?

Procedure:
Steps

1) Let a equal the first part of the ratio and let b equals the second part: $a = 3, b = 2$

2) C is going to be located $\frac{3}{5}$ of the distance between A and B

3) $\frac{3}{5}$ of the horizontal distance between A and B is added to the x coordinate of A, and $\frac{3}{5}$ of the vertical distance between A and B is added to the y coordinate of A.

4) The horizontal distance from A to B is $(6 - (-4)) = 10$, and $\frac{3}{5}(10) = 6$. Add 6 to x_1 : $-4 + 6 = 2$

5) The vertical distance from A to B is $(2 - (-1)) = 3$, and $\frac{3}{5}(3) = (9/5)$ or 1.8. Add 1.8 to y_1 : $-1 + 1.8 = 0.8$

6) The coordinates of C, are $(2, 0.8)$

7) $\overline{AC} : \overline{CB} = 3 : 2$

The steps 1-7 in Example 2 can be combined to create this formula to find the coordinates of the point that partitions a segment into a given ratio, $a : b$.

Expressing Geometric Properties with Equations

❸ **Partition by using the formula:** $\left(\left(\dfrac{a}{a+b}(x_2-x_1)\right)+x_1, \left(\dfrac{a}{a+b}(y_2-y_1)\right)+y_1\right)$

Partition $\overline{AB}$ into 2 segments in the ratio of 4:5. $A(-3, 4)$ and $B(3, -2)$.

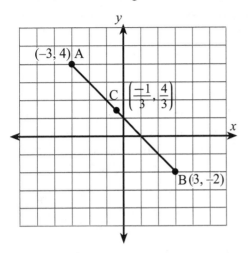

$\left(\left(\dfrac{a}{a+b}(x_2-x_1)\right)+x_1, \left(\dfrac{a}{a+b}(y_2-y_1)\right)+y_1\right)$

$\left(\left(\dfrac{4}{9}(3-(-3))\right)+(-3), \left(\dfrac{4}{9}(-2-4)+4\right)\right)$

$\left(\left(\dfrac{4}{9}\right)(6)-3, \left(\dfrac{4}{9}\right)(-6)+4\right)$

$\left(\dfrac{24}{9}-3, \dfrac{-24}{9}+4\right)$

$\left(\dfrac{24-27}{9}, \dfrac{-24+36}{9}\right)$

$\left(\dfrac{-1}{3}, \dfrac{4}{3}\right)$

The coordinates of C are $\left(\dfrac{-1}{3}, \dfrac{4}{3}\right)$. $AC : CB = 4 : 5$.

PERIMETER AND AREA

The distance formula was presented on page 126. It can be used to find the perimeter or area of polygons that are drawn on a coordinate plane. Choose the sides of the polygon that are needed to calculate the answer to the problem and use the formula to find their lengths.

$$d = \sqrt{(x_2 - x_1)^2 + (y_2 - y_1)^2}$$

Examples

❶ Lincoln Hotel is planning to reseed the grass for the rectangular lawn in front of the hotel. The corners of the land to be reseeded are shown on the following sketch. Grass seed is purchased by determining how many square feet of area it will cover. If each unit on the graph represents 50 feet, find the area that needs to be reseeded.

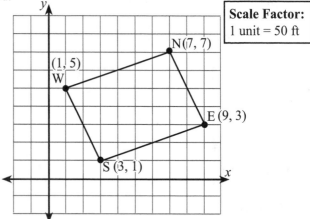

Scale Factor:
1 unit = 50 ft

Since this is a rectangle, the formula for area is $A = lw$. Find the length and the width of the rectangle and multiply each by 50. 50 is the scale factor used on the diagram above. The area of the lawn is the product of those two measurements.

$$D_{SW} = \sqrt{(3-1)^2 + (1-5)^2}$$
$$D_{SW} = \sqrt{4+16} = \sqrt{20}$$
Length of lawn $= 50\sqrt{20}$

$$D_{SE} = \sqrt{(9-3)^2 + (3-1)^2}$$
$$D_{SE} = \sqrt{36+4} = \sqrt{40}$$
Width of lawn $= 50\sqrt{40}$

$$A = lw$$
$$A = \left(50\sqrt{20}\right)\left(50\sqrt{40}\right)$$
$$A = 2500\sqrt{800}$$
$$A \approx 70,711 \, sq \, ft$$

The area of the lawn to be reseeded is about 70,711 square feet.

Expressing Geometric Properties with Equations

❷ Lincoln Hotel, in Example 1 above, wants to make a flower border all the way around the lawn. The flower border will be 1 foot wide on all sides of the lawn. How many square feet of flowers will be needed?

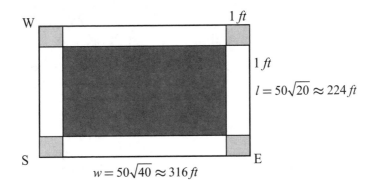

The perimeter of the lawn is needed. The length and width are already known from example 1 above. The formula for the perimeter of a rectangle is $P = 2l + 2w$. Don't forget the corners of the flower border as they are not part of the lawn measurement. Each corner is 1 ft × 1 ft.

$P_{lawn} = 2(224) + 2(316)$

$P = 1080 \, ft.$

The flower border is 1 foot wide. 1080 square feet of border is needed to edge the grass, plus the 4 square feet for the corners.

Conclusion: About 1084 square feet of flower border is needed.

Expressing Geometric Properties with Equations

❸ Find the area and perimeter of triangle ABC with vertices at $A(-3, -1)$, $B(-1, 3)$ and $C(4, -2)$.

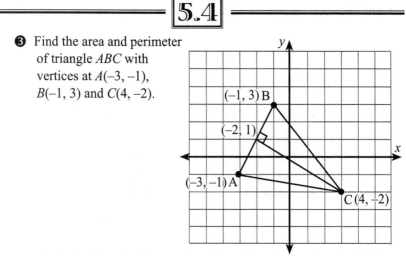

Analysis: Use the distance formula to find the length of each side and add them for the perimeter.

$$d_{BC} = \sqrt{(4+1)^2 + (-2-3)^2} = \sqrt{25+25} = \sqrt{50} = 5\sqrt{2}$$
$$d_{AC} = \sqrt{(4+3)^2 + (-2+1)^2} = \sqrt{49+1} = \sqrt{50} = 5\sqrt{2}$$
$$d_{AB} = \sqrt{(-1+3)^2 + (3+1)^2} = \sqrt{4+16} = \sqrt{20} = 2\sqrt{5}$$
$$P = 5\sqrt{2} + 5\sqrt{2} + 2\sqrt{5} = 10\sqrt{2} + 2\sqrt{5}$$

To find the area, the altitude is needed. As shown in the perimeter calculations, this is an isosceles triangle. $AC = BC$, $\overline{AB}$ is the base. The length of the altitude drawn from C to $\overline{AB}$ is needed.

Altitude: Since this is an isosceles triangle the altitude drawn from the vertex opposite the base bisects the base. Find the midpoint, D, of base $\overline{AB}$. Draw $\overline{CD}$ and find its length using the distance formula.

Midpoint Formula: $M = \left(\dfrac{x_2 + x_1}{2}, \dfrac{y_2 + y_1}{2} \right)$

Midpoint of $\overline{AB}$: $M_{AB} = \left(\dfrac{-3 + (-1)}{2}, \dfrac{-1 + 3}{2} \right)$

$$M_{AB} = (-2, 1)$$

Area of $\triangle ABC$: $A_{\triangle ABC} = \dfrac{1}{2}(2\sqrt{5})(3\sqrt{5})$

$$A = 15$$

Conclusion: The perimeter of $\triangle ABC$ is $10\sqrt{2} + 2\sqrt{5}$ units and the area is 15 sq units.

PARABOLAS

Focus and Directrix: The focus is a point inside the curve of a parabola. The parabola curved away from a line is called the directrix. The vertex of the parabola is the point on the parabola that is half the distance between the focus and the directrix along a line perpendicular to the directrix. The distance from the focus to any point on the parabola is equal to the distance between that point on the parabola and a line drawn perpendicular to the directrix

Figure 1

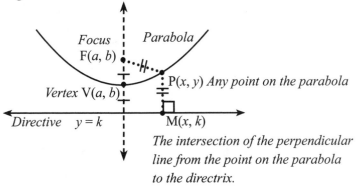

Focus F(a, b)   *Parabola*

Vertex V(a, b)   P(x, y) *Any point on the parabola*

Directive $y = k$ M(x, k)

The intersection of the perpendicular line from the point on the parabola to the directrix.

Equation of a Parabola: Two forms of a quadratic equation that will produce a parabola are:

Vertex Form: $y = r(x - a)^2 + c$, where the coordinates of the vertex are (a, c); and r is the coefficient of x^2. This labeling is shown on Figure 1. This equation is derived on the next page.

Standard Form: $y = ax^2 + bx + c$, where a is the coefficient of x^2, the squared term, b is the coefficient of x and c is a constant. This form is usually used when the quadratic formula is involved.

Note: The names of these equations and the variables used are different in various resources.

Expressing Geometric Properties with Equations

DERIVE THE EQUATION OF A PARABOLA
GIVEN A FOCUS AND DIRECTRIX

Since any point on the parabola is equidistant from the focus and the directrix along a line drawn perpendicular to the directrix, the distance formula can be used to derive an equation of a parabola. The distance between a point on the parabola to the focus, and the distance between the point on the parabola to the directrix, can be set equal to each other and the equation can be developed.

In Figure 2 a vertical parabola is given and $FP = PM$. In this case, PM is vertical, so its length is the absolute value of $y - k$.

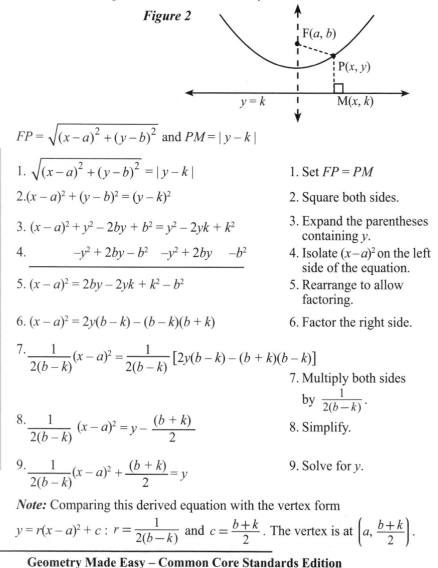

Figure 2

$FP = \sqrt{(x-a)^2 + (y-b)^2}$ and $PM = |y-k|$

1. $\sqrt{(x-a)^2 + (y-b)^2} = |y-k|$ 1. Set $FP = PM$

2. $(x-a)^2 + (y-b)^2 = (y-k)^2$ 2. Square both sides.

3. $(x-a)^2 + y^2 - 2by + b^2 = y^2 - 2yk + k^2$ 3. Expand the parentheses containing y.

4. $\underline{ -y^2 + 2by - b^2 \quad -y^2 + 2by \quad -b^2 }$ 4. Isolate $(x-a)^2$ on the left side of the equation.

5. $(x-a)^2 = 2by - 2yk + k^2 - b^2$ 5. Rearrange to allow factoring.

6. $(x-a)^2 = 2y(b-k) - (b-k)(b+k)$ 6. Factor the right side.

7. $\dfrac{1}{2(b-k)}(x-a)^2 = \dfrac{1}{2(b-k)}[2y(b-k) - (b+k)(b-k)]$ 7. Multiply both sides by $\dfrac{1}{2(b-k)}$.

8. $\dfrac{1}{2(b-k)}(x-a)^2 = y - \dfrac{(b+k)}{2}$ 8. Simplify.

9. $\dfrac{1}{2(b-k)}(x-a)^2 + \dfrac{(b+k)}{2} = y$ 9. Solve for y.

Note: Comparing this derived equation with the vertex form

$y = r(x-a)^2 + c : r = \dfrac{1}{2(b-k)}$ and $c = \dfrac{b+k}{2}$. The vertex is at $\left(a, \dfrac{b+k}{2}\right)$.

Geometry Made Easy – Common Core Standards Edition

$$\boxed{5.5}$$

❶ Find the equation of a parabola with the focus at $F(2, 1)$ and the directrix at $y = -2$.

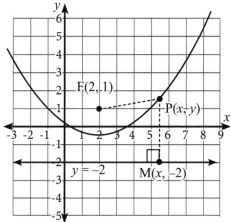

$F(2, 1)$ is the focus. The directrix is $y = -2$.

Draw a line from $P(x, y)$ perpendicular to the directrix $(y = -2)$ at point $M(x, -2)$.

The definition of a parabola tells us that $PM = FP$.
- Point P is y units above the x-axis and M is 2 units below the x-axis. The length of $PM = |y + 2|$.
- FP requires the use of the distance formula. $FP = \sqrt{(x-2)^2 + (y-1)^2}$

1. $y + 2 = \sqrt{(x-2)^2 + (y-1)^2}$

1. Set the lengths of PM and FP = to each other.

2. $(y + 2)^2 = (x - 2)^2 + (y - 1)^2$

2. Square both sides. Square the binomial containing y as indicated.

3. $y^2 + 4y + 4 = (x-2)^2 + y^2 - 2y + 1$
   $\underline{-y^2 + 2y - 4 \qquad\qquad -y^2 + 2y - 4}$
   $6y = (x-2)^2 - 3$

3. Rearrange putting y on one side, x on the other.

4. $y = \dfrac{(x-2)^2 - 3}{6}$

4. Solve for y by dividing both sides by 6.

5. $y = \dfrac{1}{6}(x-2)^2 - \dfrac{1}{2}$

5. This equation is in vertex form. Vertex is at $(2, -1/2)$.

Expressing Geometric Properties with Equations

❷ The science class was split into 2 teams to work on a robot project. The whole class designed a robot that can follow an equation programmed into it along the path indicated by the equation. One team of students designed the course as shown on the diagram below. The other team will program the robot to travel the course. The robot must start at the back of the classroom and travel around the cone and return to the back of the room. At all times the robot must be equidistant from the cone and the front wall.

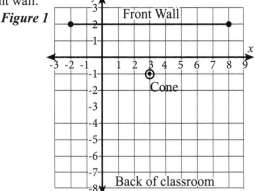

Figure 1

The team that is developing the correct equation determined that the path for the robot is a parabola. They put the diagram on a grid and labeled it. The class has worked on the equation of a parabola in their math class, and they have derived a formula that they know they can use. They label the cone as the focus at $F(3, -1)$ and the front wall line as the directrix at $y = 2$. Figure 2 shows the path the robot took to accomplish the task.

Finding the equation: Substitute $(3, -1)$ for a and b, and $k = 2$ since the directrix is at $y = 2$.

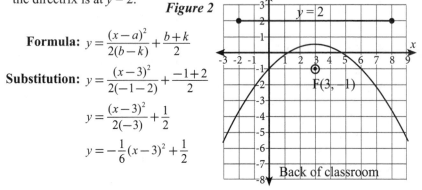

Figure 2

Formula: $y = \dfrac{(x-a)^2}{2(b-k)} + \dfrac{b+k}{2}$

Substitution: $y = \dfrac{(x-3)^2}{2(-1-2)} + \dfrac{-1+2}{2}$

$y = \dfrac{(x-3)^2}{2(-3)} + \dfrac{1}{2}$

$y = -\dfrac{1}{6}(x-3)^2 + \dfrac{1}{2}$

Programming the robot with the equation $y = -\dfrac{1}{6}(x-3)^2 + \dfrac{1}{2}$ allowed it to complete the path as the rules directed.

Unit 6

CIRCLES WITH AND WITHOUT COORDINATES

- Understand formulas for circles.

- Understand and apply theorems about circles.

- Find arc lengths and areas of sectors of circles.

- Translate between the geometric description and the equation for a conic section.

- Use coordinates to prove simple geometric theorems algebraically.

- Apply geometric concepts in modeling situations.

CIRCUMFERENCE

The formula for the circumference of a circle $C = 2\pi r$ is already familiar. To understand why this works, consider the angles and sides of a polygon inscribed in a circle. As the number of sides of an inscribed regular polygon increases, the perimeter of the polygon is closer and closer to being equal to the circumference of the circle containing it.

Example

Given: Circle O with radius = 1.

Prove: $C = 2\pi r$

Steps:

1) Inscribe an equilateral triangle, ABC, in Circle O. Circle O has radius = 1.

2) Connect O with A, B, and C.
3 $\cong$ isosceles triangles are formed.
Use $\triangle BOC$: The congruent sides = $r = 1$, the base is $\overline{BC}$.

3) Draw altitude, h, from O to $\overline{BC}$, label the intersection D.
$\overline{OD}$ bisects $\overline{BC}$ and $\angle BOC$.

4) Central $\angle BOC = \dfrac{360°}{3} = 120°$; $\angle BOD = \dfrac{1}{2}(120)° = 60°$

5) Use right triangle trig to find the length of $\overline{DC}$, then multiply by 2 to find the length of $\overline{BC}$. $\overline{BC}$ is the length of one side, s, of the inscribed polygon.

$$Sin\ 60 = \frac{DC}{r} \qquad 2(DC) = BC$$

$$r\ Sin\ 60 = DC \qquad BC \text{ or } s \approx 2(0.8660254038) \approx 1.732050808$$

$$(1)Sin\ 60 = DC$$

$$DC \approx 0.8660254038$$

6) Multiply $\overline{BC}$, or s, by 3 to find the perimeter of $\triangle ABC$, the inscribed polygon.

$$P \approx 3(1.732)$$

$$P \approx 5.196152423$$

Note: In circle O with $r = 1$, the approximate circumference is 2π or about 6.28. More work is needed since approximately 5.196 is not a very accurate estimate.

Geometry Made Easy – Common Core Standards Edition

Use inscribed polygons with more sides to continue the proof.

7) Continue to do the same process with polygons that have increasing numbers of size inscribed in a circle with radius = 1.

8) Notice what happens to h and r as the number of sides in the polygon increases and the side, s, becomes smaller.

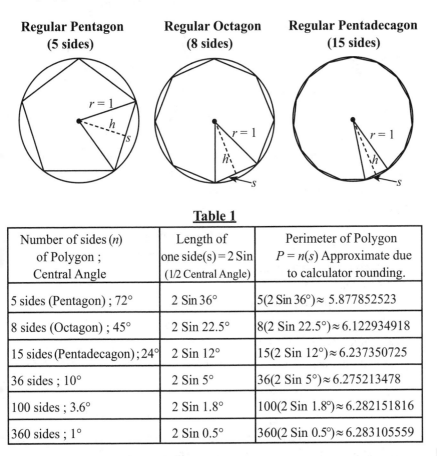

Regular Pentagon (5 sides)	Regular Octagon (8 sides)	Regular Pentadecagon (15 sides)

Table 1

Number of sides (n) of Polygon ; Central Angle	Length of one side(s) = 2 Sin (1/2 Central Angle)	Perimeter of Polygon $P = n(s)$ Approximate due to calculator rounding.
5 sides (Pentagon) ; 72°	2 Sin 36°	5(2 Sin 36°) ≈ 5.877852523
8 sides (Octagon) ; 45°	2 Sin 22.5°	8(2 Sin 22.5°) ≈ 6.122934918
15 sides (Pentadecagon);24°	2 Sin 12°	15(2 Sin 12°) ≈ 6.237350725
36 sides ; 10°	2 Sin 5°	36(2 Sin 5°) ≈ 6.275213478
100 sides ; 3.6°	2 Sin 1.8°	100(2 Sin 1.8°) ≈ 6.282151816
360 sides ; 1°	2 Sin 0.5°	360(2 Sin 0.5°) ≈ 6.283105559

Conclusion: As the number of sides of the regular polygon increases, the sides of the polygon become closer and closer to the circle itself. The perimeter can never become larger than the circle and still be inscribed. The perimeter approaches the value of 2π which is approximately 6.283185307. Since the radius is one, the actual circumference of these circles is $C = 2\pi(1)$.

The actual circumference of Circle O : $C = 2\pi(1)$ *or* 2π.

Circles

AREA OF A CIRCLE

The same thought process that was used to justify the formula for the circumference of the circle can be used to justify the formula for the area of a circle, $A = \pi r^2$. Instead of using the perimeter of the polygon, the area of the polygon will show the results needed.

Note: The numbers are rounded in the examples, causing the answers to be approximate.

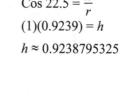

Example

Given: Circle O with radius = 1.

Prove: $A = \pi r^2$

Steps:

1) Inscribe a regular polygon. Find the length of each side of the polygon as shown in steps 1-5 on the previous page. This is the base of the isosceles triangle needed to find area.

2) Using the cosine function, find the length of the altitude.

The side of the inscribed polygon is the base, b, of the triangle.

See also page 140–141.

$$\text{Cos } 22.5 = \frac{h}{r}$$

$$(1)(0.9239) = h$$

$$h \approx 0.9238795325$$

3) Find the area of one triangle and multiply by n, the number of sides in the polygon.

$$A = \frac{1}{2}bh$$

$$A \approx \frac{1}{2}(0.7653668647)(0.9238795325)$$

$$A \approx 0.3536$$

$$8A \approx 8(0.3536) = 2.8286$$

4) Continue with several more polygons to make a table showing the area of the polygons. A summary of this is shown in Table 2.

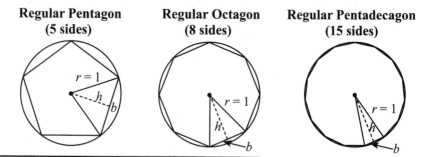

| Regular Pentagon (5 sides) | Regular Octagon (8 sides) | Regular Pentadecagon (15 sides) |

Circles

Number of Sides (n) ; Central Angles	Length of one side (b) = 2 Sin ½ Cnt. Angle	Altitude (h) = Cos ½ Cnt. Angle	Area of Polygon = n (½) bh (Approximation due to calculator rounding.)
5 (Pentagon) ; 72°	2 Sin 36°	Cos 36°	5(0.5)(2 Sin 36°)(Cos 36°) ≈ 2.377641291
8 (Octagon) ; 45°	2 Sin 22.5°	Cos 22.5°	8(0.5)(2 Sin 22.5°)(Cos 22.5°)≈2.828427125
15 (Pentadecagon);24°	2 Sin 12°	Cos 12°	15(0.5)(2 Sin 12°)(Cos 12°) ≈ 3.050524823
36 sides ; 10°	2 Sin 5°	Cos 5°	36(0.5)(2 Sin 5°)(Cos 5°) ≈ 3.1256674198
100 sides ; 3.6°	2 Sin 1.8°	Cos 1.8°	100(0.5)(2 Sin 1.8°)(Cos 1.8°)≈3.139525976
360 sides ; 1°	2 Sin 0.5°	Cos 0.5°	8(0.5)(2 Sin 22.5°)(Cos 22.5°)≈3.141433159

*Numbers are approximate due to rounding.

Conclusion: As the number of sides of the regular polygon increases, the area of the polygon is closer and closer to filling the circle completely. The area of the polygon approaches 3.14159…. which is the approximate value of π. In Table 2 above, the radius is 1, making the area of the circle equal to $\pi(1)^2$ or the value of π, approximately 3.14.

Option: Show what happens using an inscribed pentadecagon (15 sides) if the radius is 3 instead of 1. The numbers here are rounded for convenience.

Length of Side (b):

$$Sin\ 12 = \frac{0.5b}{3}$$

$$\frac{3\,Sin\,12}{0.5} = b$$

$$b \approx 1.247$$

Altitude (h)

$$Cos\ 12 = \frac{h}{3}$$

$$3Cos\ 12 = h$$

$$h \approx 2.934$$

Area of Polygon:

$$A = \frac{1}{2}bh(n)$$

$$A \approx \frac{1}{2}(1.247)(2.934)(15)$$

$$A \approx 27.440$$

Actual Area of the Circle with $r = 3$:

$A = \pi r^2$

$A = 9\pi$

$A \approx 28.274$

Conclusion: When the radius of the circle increases, the relationship of $A = \pi r^2$ remains true.

Discussion: What happens to the circumference formula if the radius increases? What will the change be in the area formula and/or the circumference formula if the radius is less than 1?

Answer: There will be no change in the formula for either. $A = \pi r^2$

Circles

CIRCLES AND ANGLES

Circle: The set of points a given distance (called the radius) from a point called the center. A circle is measured in degrees. There are 360° in a circle. A circle is named by its center point.

Radius: A line segment from any point on the circle to the center.

Diameter: A line segment from any point on the circle that goes through the center to another point on the circle.

Circumference: The distance around the edge of the circle.
  Formula: $C = 2\pi r$ or $C = \pi d$

Area: The space enclosed by or inside the circle.   Formula: $A = \pi r^2$

Subtend: When the end points of an arc are formed by the intersection of the rays of an angle and the circle, the rays subtend (or cut off) the arc. $\angle CAD$ subtends $\overgroup{CD}$.

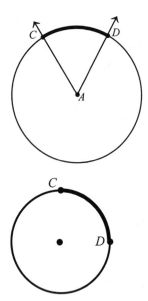

Sector: The portion of a circle bordered by two radii and an arc between them.

Arc: Part of the circumference. Arc CD is written $\overgroup{CD}$. It means the part of the circumference that is between points C and D. It can be measured in degrees, and it can be measured in linear measure called the arc length.

Locating the Arc: Any arc can be labeled with multiple points – depending on what points are labeled on the circle between the endpoints of the arc. When locating the arc, start on the first letter named and move toward the next letter named in the shortest way. Continue until you get to the last letter named.

Circles

TYPES OF ARCS

1. **Minor Arc:** An arc that measures < 180°.
   It is labeled with 2 letters. $\overset{\frown}{BC}$ in Figure 1.

2. **Major Arc:** An arc that measures > 180°.
   It is usually labeled with 3 letters. $\overset{\frown}{CDF}$ in Figure 1.

3. **Semi-circle:** An arc that measures 180°. Its points on the circle are
   the endpoints of a diameter of the circle. $\overset{\frown}{FCD}$ in Figure 1.

Example

Figure 1

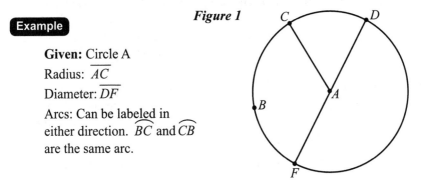

Given: Circle A

Radius: $\overline{AC}$

Diameter: $\overline{DF}$

Arcs: Can be labeled in
either direction. $\overset{\frown}{BC}$ and $\overset{\frown}{CB}$
are the same arc.

Identify the three types of arcs on the circle on Figure 1 and label

1. **Minor Arcs:** $\overset{\frown}{BC}$, $\overset{\frown}{BD}$ or $\overset{\frown}{BCD}$, $\overset{\frown}{CD}$, $\overset{\frown}{BF}$, $\overset{\frown}{CBF}$ or $\overset{\frown}{CF}$
   are all minor arcs. They each measure < 180°.

2. **Major Arcs:** $\overset{\frown}{BCF}$ or $\overset{\frown}{BCDF}$, $\overset{\frown}{CDF}$, $\overset{\frown}{CDFB}$ or $\overset{\frown}{CFB}$
   are all major arcs. Each measures > 180°.

3. **Semi Circles:** $\overset{\frown}{DF}$ or $\overset{\frown}{FD}$, $\overset{\frown}{FBCD}$, $\overset{\frown}{DCBF}$. These are ways
   to name the semicircles. Each measures 180°.

Circles

CHORDS, TANGENTS AND SECANTS

Chords, tangents, and secants are the types of lines and segments that are used with circles.

Chord: A segment that connects two points on a circle. It is in the interior of the circle.

Tangent: A line that extends from a point outside the circle to the circle, touching the circle at exactly one point. Tangent $\overline{IJ}$ is tangent to circle A at M in Figure 2 below. It does not enter the circle. If a tangent is extended past the circle, its extension goes "past" the circle, not into it. We often use part of a tangent as a segment ($\overline{IM}$ or $\overline{JM}$ in Figure 2 below).

Secant: A line that passes through two points on a circle. We often use part of a secant as a segment. It connects a point outside the circle, intersects the circle and goes across it to another point on the other side of the circle. (See $\overline{LR}$ in Figure 2.)

Chord: $\overline{BH}$ Another chord is $\overline{RQ}$ which is part of the secant $\overline{RL}$.

Tangent: $\overline{IJ}$ intersects circle A at M.

Secant: $\overline{RL}$ which intersects circle A at R and Q.

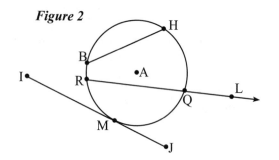

Figure 2

Circles

SIMILARITY OF CIRCLES

Proving all circles are similar seems pretty easy since they look alike except for size. But in order for a figure to be similar to another, one or more transformations must be able to convert one figure to the other.

Examples

Figure 1

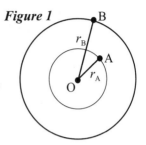

❶ The circles in this example are named A and B for clarity. Point O is the center of both circles. $r_{OB} = 2r_{OA}$

1. Circle A with radius $\overline{OA}$, is dilated by a scale factor of 2 with the center of dilation at O.

2. Circle B with radius $\overline{OB}$, is the image of circle A after the dilation.

3. $\dfrac{2r}{r} = 2$. 2 is the constant of dilation, or the scale factor.

4. A dilation is a non-rigid motion in which similarity is preserved, making the two circles similar.

❷ Compare the circumferences of the 2 circles in Figure 1. If they are similar, their circumferences should be in the same proportion as their radii.

$C = 2\pi r$
$C_A = 2\pi r$
$C_B = 2\pi(2r)$
$\dfrac{C_B}{C_A} = \dfrac{4\pi r}{2\pi r} = 2$

The scale factor between the circumference of circle B to circle A is 2, which is the same scale factor as was used to dilate the radius when forming circle B.

❸ Given circles A, A' and A''. Prove that they are similar.

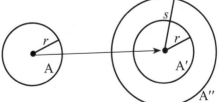

Paragraph Proof: Given are 3 circles, A, A' and A''. Circle A' is a translation of Circle A along the vector shown. $\odot A \sim \odot A'$ because a translation is a rigid motion and the image is both congruent and similar to the pre-image. Circle A'' is a dilation of Circle A' that maps it to Circle A'' using a scale factor of $\frac{s}{r}$. $\odot A' \sim \odot A''$ since a dilation is a non-rigid motion transformation and preserves similarity. $\odot A \sim \odot A''$ using the transitive property. Since $\odot A \sim \odot A'$ and $\odot A' \sim \odot A''$ then $\odot A \sim \odot A''$.

Conclusion: All of the circles are similar.

Given: Circle A.
Prove: Circle A is similar to circle A''.

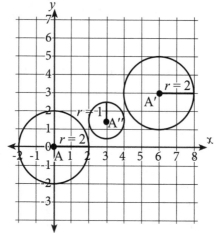

Steps	Explanation
1. $\odot A \xrightarrow{T_{6,3}} \odot A'$ $\odot A \cong \odot A', r_{A'} = r_A$	1. Translate circle A to circle A'. $\odot A \cong \odot A'$ because a translation is a rigid motion transformation.
2. $\odot A' \xrightarrow{D_{0.5}} \odot A''$	2. Dilate A' using a scale factor of 0.5. The center of dilation in this example is the origin. The scale factor can be found by comparing the radius of A' with A''.
3. $\odot A \xrightarrow{D_{0.5} \circ T_{6,3}} \odot A''$	3. A composite transformation is performed on circle A resulting in similarity of the pre-image and the image.
4. $\odot A \sim \odot A''$	4. Both transformations result in images that are similar to the pre-image.

CIRCLES & THEIR ANGLES & ARCS

ANGLES AND ARCS

The angles formed by the segments and lines associated with a circle have rules for finding their measure that are related to the arcs they intercept or subtend. Some have special names, others are named descriptively. If the measures of the arcs are available, the measures of the related angles can be found. And conversely, if the measures of the angles are available, the measures of the arcs can be found.

SUM OF THE ARCS

The sum of the arcs in any circle is 360°. Ratios are sometimes used to find the degree measure of the arcs in a circle.

Examples

❶ In circle O, the ratio of $\overset{\frown}{ABC}$ to $\overset{\frown}{AC}$ is 3:2. Find the degree measure of each arc.

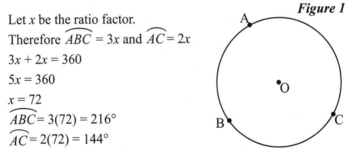

Figure 1

Let x be the ratio factor.
Therefore $\overset{\frown}{ABC} = 3x$ and $\overset{\frown}{AC} = 2x$

$3x + 2x = 360$

$5x = 360$

$x = 72$

$\overset{\frown}{ABC} = 3(72) = 216°$

$\overset{\frown}{AC} = 2(72) = 144°$

❷ Arcs $\overset{\frown}{ABC}, \overset{\frown}{CD}, \overset{\frown}{DA}$ and are in a ratio of 5:2:3 respectively. Find the measure of each arc.

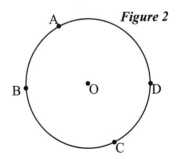

Figure 2

Let $5x = \overset{\frown}{ABC}$
Therefore $2x = \overset{\frown}{CD}, 3x = \overset{\frown}{DA}$

$2x + 3x + 5x = 360$

$10x = 360$

$x = 36$

$\overset{\frown}{ABC} = 5(36) = 180°$

$\overset{\frown}{CD} = 2(36) = 72°$

$\overset{\frown}{DA} = 3(36) = 108°$

Circles

INTERIOR ANGLES

Central Angle: The vertex is at the center of the circle, and both sides are radii. The intercepted arc equals the degree measure of the central angle. The radii of a central angle subtend an arc creating a sector of the circle. *POQ* is a sector created by central angle *O* and its radii, $\overline{OQ}$ and $\overline{OP}$. (See Figure 3)

Inscribed Angle: The vertex is on the circle and the sides are chords (Figure 3), a chord and a secant (Figure 4), or 2 secants. The measure of the inscribed angle $= \frac{1}{2}$ the measure of the arc it intercepts.
See Figure 3 and Figure 4.

Examples

> **Given:** Circle *O*
>
> Radii $\overline{OQ}$ and $\overline{OP}$
>
> Chords $\overline{ST}$ and $\overline{SR}$
>
> Arc measures are labeled.

❶ **Central Angle:** $m\angle QOP = m\overset{\frown}{QP}$; $m\angle QOP = 46$; $m\overset{\frown}{QP} = 46$

Note: A central angle is the <u>only</u> angle in the group of angles related to circles that is equal to the arc it cuts off.

Note: A diameter forms a central angle of 180° since it intercepts $\frac{1}{2}$ of the 360° circle.

❷ **Inscribed Angle:** $m\angle TSR = \frac{1}{2}\overset{\frown}{TR}$; $m\angle TSR = \frac{1}{2}(136) = 68$

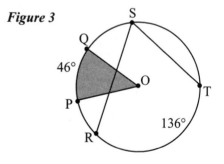

Figure 3

Circles

Other angles with a vertex on the circle are not inscribed angles. These angles are formed by a tangent and a secant ($\angle 1$ in Figure 4) or a tangent and a chord ($\angle 2$ in Figure 4): The vertex is on the circle, one side is a tangent and the other is a chord or secant. The measure of this angle $= \dfrac{1}{2}$ its intercepted arc.

Figure 4

Given: Circle A

Tangent $\overline{CED}$

Chord $\overline{EF}$

Secant $\overline{EG}$

Arc measures are labeled.

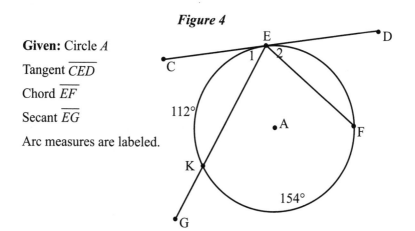

$m\angle 1 = \dfrac{1}{2}\, m\,\overset{\frown}{EK}$; $m\angle 1 = \dfrac{1}{2}(112) = 56°$

$m\angle 2 = \dfrac{1}{2}\, m\,\overset{\frown}{EF}$; $m\angle 2 = \dfrac{1}{2}(94) = 47°$

Find $m\,\overset{\frown}{EF}$ first: $360 - (112 + 154) = 94°$

Inscribed $\angle KEF = \dfrac{1}{2}\, m\overset{\frown}{KF}$; $m\angle KEF = \dfrac{1}{2}(154) = 77°$

- *It doesn't matter whether the sides are tangents, chords, or secants, the same rule applies. When the vertex is ON the circle, the angle measures $\dfrac{1}{2}$ the measure of its intercepted arc.*

Circles

INTERIOR ANGLES FORMED BY TWO CHORDS

Two chords that intersect in a circle: An angle formed by two chords intersecting inside a circle is equal to half the sum of the arc it intercepts and the arc its vertical angle intercepts.

Given: Circle A,

chords $\overline{FE}$ and $\overline{CD}$.

Arc measures are labeled.

Figure 6

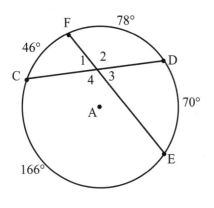

$\angle 1$ and $\angle 3$ are vertical angles.

$$m\angle 1 = \frac{1}{2}(m\,\overset{\frown}{CF} + m\,\overset{\frown}{DE})$$

$$m\angle 1 = \frac{1}{2}(46 + 70) = 58°$$

$$m\angle 3 = 58°$$

2 and 4 are vertical angles.

$$m\angle 2 = \frac{1}{2}(m\,\overset{\frown}{FD} + m\,\overset{\frown}{CE})$$

$$m\angle 2 = \frac{1}{2}(78 + 166) = 122°$$

$$m\angle 4 = 122°$$

Note: Remember this rule is a sum because the intersection of the two chords inside the circle looks something like a plus sign.

An interior angle formed by two chords is the only type of angle that equals $\frac{1}{2}$ the SUM of its arc and the arc of its vertical angle.

Circles

ANGLES OF SECTORS

Radian Measure: Angles and arcs of a circle are often measured in radians instead of degrees. The degree measure of a circle is 360°, and the radian measure is $2\pi r$. Either can be used to describe the total angle in one rotation of a radius about the center of the circle, and also to measure the distance around the circle that is measured by one complete rotation –the circumference. Since $2\pi r$ and 360° measure the same things, they can be made equal to each other. The radian measure of $2\pi r$ equals the degree measure of 360°.

RADIANS AND DEGREES

Angles are measured in degrees and also in radians. (The calculator has two modes – degrees and radians.) One radian is the measure of a central angle that subtends (intersects) an arc measured along the circumference of the circle that is equal in length to the radius of the circle. Angles measured in degrees will be labeled with a degree mark: °. Angles measured in radians do not always have a label – they may contain the π symbol or they may be "just a number".

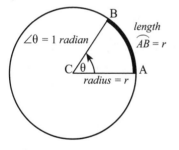

Since the circumference of a circle is equal to $2\pi r$, the radius can be measured along the circumference 2π times. There are 2π radians in one complete rotation of the circle. Remember there are 360° in one complete rotation as well.

360° = 2π radians

Constant of Proportionality: As with other similar figures, a constant of proportionality can be applied to circles. In circles with the same radii, the ratio of the central angles is the constant of proportionality used to determine the arc length of the sectors. In circles with different length radii, but equal central angles, the ratio of the radii is the constant of proportionality used to find the arc lengths.

Circles

$$\boxed{6.5}$$

Changing from Degrees to Radians: Use the conversion factor $\dfrac{\pi \text{ radians}}{180°}$ and multiply the degrees by the conversion factor. Reduce the fraction if possible. The radian measure is usually a simplified fraction left in terms of π.

Examples

❶ $45° \cdot \dfrac{\pi}{180°} = \dfrac{45\pi}{180} = \dfrac{\pi}{4}$ radians

❷ $150° \cdot \dfrac{\pi}{180°} = \dfrac{150\pi}{180} = \dfrac{5\pi}{6}$ radians

Changing from Radians to Degrees: Use the conversion factor $\dfrac{180°}{\pi \text{ radians}}$. Multiply the radians by the conversion factor.

Examples

❶ $\theta = \dfrac{5\pi}{3} \cdot \dfrac{180°}{\pi} = 300°$

❷ $\theta = 1.2 \text{ radians}$

$\theta = 1.2 \cdot \dfrac{180°}{\pi} = 68.75°$ (rounded to the *hundredths place*)

❸ In circle O, the radius is 3. The length of arc BC is 6 cm. What is the measure of θ ($\angle BOC$) in radians and in degrees?

In radian measure: $\overarc{BC} = 6$,

$r = 3$ $\overarc{BC}$ is twice the radius, $\therefore \theta = 2$ radians

In degree measure: $\theta = 2\left(\dfrac{180°}{\pi}\right) \approx 114.6°$

(rounded to the *tenths place*)

ARC LENGTH AND AREA OF SECTORS

To discuss the arc lengths and areas of sectors, radian measure of the central angle is used. Changes in the length of the radius or in the measure of the central angle, θ, have an effect on the length of the subtended arc or the area of the sector. A sector is labeled using its two endpoints on the circle with the center of the circle. The arc length or area of a sector is a part of the entire circumference or the entire area respectively.

$$\frac{m°}{360°} = \frac{\theta}{2\pi} = \frac{Arc_{length}}{2\pi r} \qquad \frac{m°}{360°} = \frac{\theta}{2\pi} = \frac{Arc_{sector}}{\pi r^2}$$

ARC LENGTH OF A SECTOR

The arc length of a sector is found using the following formula:
$s = r\theta$ *when is θ measured in radians.*

Examples

❶ Demonstrate the procedure(s) that can be used to find the length of an arc that is subtended by a central angle of 90° in a circle with a radius equal to 1 cm.

Plan: Determine what fraction of the total circle is measured by the central angle? The central angle is 90°. 90° is $\frac{1}{4}$ of the total of 360° in the circle. Therefore the arc should be $\frac{1}{4}$ of the circumference. Change the degree measure to radian measure.

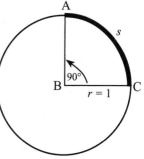

Steps

1) Convert to radians $(90°) \bullet \left(\frac{\pi}{180°}\right) = \frac{\pi}{2}; \ \theta = \frac{\pi}{2}$

2) Formula $s = (\theta) \bullet (r)$

3) Substitute $s = \left(\frac{\pi}{2}\right) \bullet (1)$

4) Simplify $s = \frac{\pi}{2} cm$ or $s \approx 1.57 cm$

Discussion: Approximately 6.28 or 2π is the circumference of the circle. Since $1.57 \approx \frac{1}{4}(6.28)$ and $\frac{\pi}{2} = \left(\frac{1}{4}\right) 2\pi$, the answer makes sense.

Conclusion: Since 1 radian is equivalent to the length of the radius, if the radius is 1, the subtended arc is equal to the value of the central angle in radians. In this example, the angle is $\frac{\pi}{2}$ and the subtended arc is the same.

Circles

❷ Find the length of $\overset{\frown}{AC}$ if $\angle ABC = 60°$, $r = 1$ cm. (Notice the radius in this example is the same as in Example 1 on page 157.)

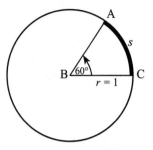

Convert to radians $s = 60° \bullet \dfrac{\pi}{180°} = \dfrac{\pi}{3}$

$$s = \dfrac{\pi}{3} \bullet 1$$

$$s = \dfrac{\pi}{3} \, cm$$

When θ changes, what results can be expected?

Compare the results of Example 1 and Example 2. The radii in both examples are the same, but the arc length in example 1 is $\dfrac{\pi}{2}$ and the arc length in example 2 is $\dfrac{\pi}{3}$. The arc length changes because the radian value of θ changes.

The ratio of the values of θ is $\dfrac{Ex1}{Ex2} : \dfrac{\frac{\pi}{2}}{\frac{\pi}{3}} = \dfrac{\pi}{2} \bullet \dfrac{3}{\pi} = \dfrac{3}{2}$

The ratio of the arc lengths $\dfrac{Ex1}{Ex2} : \dfrac{\frac{\pi}{2}}{\frac{\pi}{3}} = \dfrac{\pi}{2} \bullet \dfrac{3}{\pi} = \dfrac{3}{2}$

The constant of proportionality is $\dfrac{3}{2}$.

Conclusion: In two circles, when the measure of the central angle changes and the radii are the same, the arc lengths change in the same proportion as the central angles. The constant of proportionality equals the ratio of the measures of the angles.

What happens when the radius changes? (See Example 3)

Circles

---

❸ Find the length of $\overset{\frown}{AC}$ if the central angle is $\frac{\pi}{3}$ in radian measure. The radius is 2 feet. Compare and discuss the results of Example 3 with the results of Example 2.

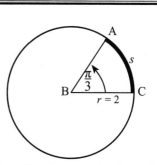

Analysis: The central angle is already in radian form, so the formula can be used directly. Notice that the radius in this example is twice the radius of Example 2 although the central angle is the same.

$s = r\theta$

$s = 2\left(\dfrac{\pi}{3}\right)$

$s = \dfrac{2\pi}{3}\ feet$

Comparison: The length of the arc in Example 3 is twice the length of the arc length in Example 2. When the radius was multiplied by two, the arc length was also multiplied by 2.

Discussion: Since we know that all circles are similar, their radii are proportional. In circles with different radii, if the central angles are equal in two circles respectively, their radii are proportional. The arc length of the 2 sectors will be proportional as well in the same scale factor or constant of proportionality as the radii.

Comparison: The arc lengths of Example 2 to Example 3 have a ratio of $\frac{1}{2}$. Their central angles are equal. The constant of proportionality (scale factor) of the radii is $\frac{1}{2}$.

$\dfrac{r_1}{r_2} = \dfrac{1}{2}$

$\dfrac{s_1}{s_2} = \dfrac{\frac{\pi}{3}}{\frac{2\pi}{3}} = \dfrac{1}{2}$

Conclusion: In circles with different radii, when the central angles are equal, the arcs of two sectors have the same proportionality as their radii. The ratio of the radii is the same as the ratio of the arc lengths.

AREA OF A SECTOR

As explained in Arc Length, the central angle of a sector is used when finding the area of a sector. The same type of reasoning is used, considering what part of the complete circle is in the sector compared with the area of the entire circle.

$$\frac{m°}{360°} = \frac{\theta}{2\pi} = \frac{A_{sector}}{A_{circle}}$$

These examples will demonstrate the development of the formula for the area of a sector.

Examples

❶ In circle B, the central angle $B = 120°$. The radius is 10 cm. Find the area of sector ABC.

Steps

1) Find the fraction of the 360° circle that is included in $\angle B$.

$$\frac{120°}{360°} = \frac{1}{3}$$

2) Find the area of Circle B.

$A = \pi r^2$
$A = (10)^2$
$A = 100\pi$

3) In this example, $\frac{1}{3}$ of the area of the circle is in sector ABC.

$A_{sector} = \frac{1}{3}(100\pi)$
$A_{sector} = 33\frac{1}{3}\pi \approx 104.72 \ cm^2$

❷ Given a circle with a radius of 5 cm and a sector of the circle with a central angle measuring $\frac{\pi}{4}$. Find the area of the sector to the nearest hundredth.

Steps

1) Determine what fraction of the 360° circle is contained in the central angle.

$$\frac{\frac{\pi}{4}}{2\pi} = \frac{\pi}{4} \cdot \frac{1}{2\pi} = \frac{1}{8}$$

2) Find the area of the entire circle:

$A = \pi r^2$
$A = (5)^2$
$A = 25\pi$

3) In this example $\frac{1}{8}$ of the area 25π is the area of the sector.

$A_{sector} = \frac{1}{8}(25\pi) \approx 9.82 \ cm^2$

Conclusion: In a sector where the central angle is given in degrees, $m°$, the area of the sector can be found by finding the portion of the entire circle contained in the sector.

The Formula for the area of a sector: $A_{sector} = \left(\frac{m°}{360°}\right)\pi r^2 \quad or \quad \left(\frac{\theta}{2\pi}\right)(\pi r^2)$

Circles

EXTERIOR ANGLES

The vertex is away from the circle. It is formed by tangents and secants: The measure of the angle equals $\frac{1}{2}$ the difference of the arcs it intercepts.

1. External angle formed by 2 tangents: The vertex is outside the circle. Both sides are tangents. $\angle BCE$ in Figure 5. See Example 1.

2. External angle formed by a secant and a tangent: $\angle BCD$ and $\angle DCE$ in Figure 5. See Example 2.

3. Exterior angle formed by 2 secants. $\angle WFD$ in Figure 5. See Example 3.

Examples

❶ $m\angle BCE = \frac{1}{2}(m\stackrel{\frown}{BWE} - m\stackrel{\frown}{BJE})$

$m\stackrel{\frown}{BWE} = 45 + 24 + 52 + 82 + 52 = 255$

$m\stackrel{\frown}{BJE} = 75 + 30 = 105$

$m\angle BCE = \frac{1}{2}(255 - 105) = 75$

Figure 5

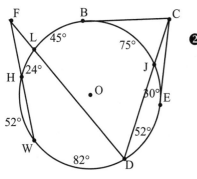

❷ Two different examples are shown here.

1. $m\angle BCD = \frac{1}{2}(m\stackrel{\frown}{BWD} - m\stackrel{\frown}{BJ})$

$m\angle BCD = \frac{1}{2}(203 - 75) = 64°$

2. $m\angle DCE = \frac{1}{2}(m\stackrel{\frown}{DE} - m\stackrel{\frown}{JE})$

$m\angle DCE = \frac{1}{2}(52 - 30) = 11°$

❸ $m\angle WFD = \frac{1}{2}(m\stackrel{\frown}{WD} - m\stackrel{\frown}{HL})$; $m\angle WFD = \frac{1}{2}(82 - 24) = 29°$

Circles

Note: In this group of angles, each one has its vertex outside or "away" from the circle. If you remember that "take <u>away</u>" means subtract, it might help you remember that you have to subtract the smaller arc from the larger one and then divide by 2 for each of these angles.

• *When the vertex is external (away from the circle), its measure*
  is $\frac{1}{2}$ the difference of its intercepted arcs.

CIRCLES AND SEGMENTS

CIRCLES AND SEGMENTS

Finding the length of various segments of chords, secants, and tangents that intersect each other in a given circle requires that different formulas be used to calculate the lengths. These problems sometimes result in a quadratic equation that must be solved.

- Two chords that intersect in a given circle: The product of the segments of one chord is equal to the product of the segments of the other.

  In circle A, chords $\overline{BE}$ and $\overline{CD}$ intersect at F. (See Figure 7)
  Rule: $(BF)(FE) = (CF)(FD)$

Examples

❶ If $BF = 5$, $CF = 7$, and $FD = 6$, find FE.
Let $x = FE$
$5x = 7(6)$
$x = 8\dfrac{2}{5}$ or 8.4
$FE = 8.4$

Figure 7

❷ **Given:** $CF = FD$, $FB = 2$,
$FE = 12$ more than CF.
Find the length of FE.
Let $x = CF$ and FD
$\therefore x + 12 = FE$
$(x)(x) = 2(x + 12)$
$x^2 = 2x + 24$
$x^2 - 2x - 24 = 0$
$(x - 6)(x + 4) = 0$
$x = 6, \ x = -4$ reject
$FE = x + 12 = 6 + 12 = 18$

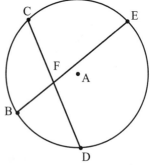

- Two tangents that share one external point: When two tangents share the same external point, the segments from that point to the points of tangency on the circle are equal.

Example

Figure 8

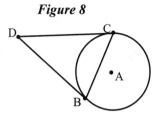

Given: Circle *A* with tangents $\overline{DB}$ and $\overline{DC}$ that intersect at *D*. Chord $\overline{BC}$.

Prove: $\triangle BDC$ is isosceles.

Solution:

Circle Proof in paragraph form: Since we are given two tangents, $\overline{DB}$ and $\overline{DC}$, that intersect at *D*, we know that $\overline{DB} \cong \overline{DC}$ because two tangents that share the same external point are congruent (or equal). That makes two sides of $\triangle BDC$ congruent. Therefore, $\triangle BDC$ is isosceles because the definition of an isosceles triangle is that two sides must be congruent.

- Two secants that intersect at the same external point: When two secants are drawn from the same external point to a circle; the product of the external segment of one secant and its entire length is equal to the product of the external segment of the other and its entire length. In the example below, $(DE)(DC) = (DF)(DB)$.

Example

If $CE = 2$, $DE = 4$, and $BF = 5$, find DF.

Figure 9

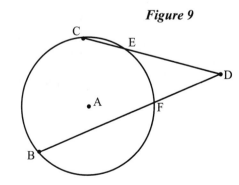

Let x = DF

$(4)(4 + 2) = x(x + 5)$

$24 = x^2 + 5x$

$x^2 + 5x - 24 = 0$

$(x + 8)(x - 3) = 0$

$x = -8$ reject, $x = 3$

$DF = 3$

Circles

- A tangent and a secant intersect at the same external point: The length of the tangent is the mean proportional between the external segment of the secant and the entire secant. The same situation can be described this way: The tangent squared is equal to the product of the external segment of the secant and the entire secant.

Example $\dfrac{DE}{CD} = \dfrac{CD}{BD}$ or $(CD)^2 = (BD)(DE)$

If $BE = 9$ and $DE = 3$, find CD.
Two solutions are shown.

Let x = CD

Method 1	Method 2
$\dfrac{ED}{CD} = \dfrac{CD}{(BE + ED)}$	$(CD)^2 = (ED)(BE + ED)$
$\dfrac{3}{x} = \dfrac{x}{(9 + 3)}$	$x^2 = (3)(9 + 3)$
$x^2 = 36$	$x^2 = 36$
$x = \pm\,6$, reject -6	$x = 6$
$CD = 6$	$CD = 6$

Figure 10

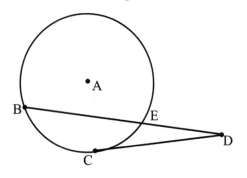

Circles

CHORDS, TANGENTS, RADII, DIAMETERS, AND ARCS

These are additional statements that can be used in solving circles or doing circle proofs. "Solving" a circle means to find as much information as possible about its arc, angles, and segments.

1. If a radius is perpendicular to a chord, it bisects the chord and its arc.

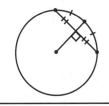

2. If a diameter or radius meets a tangent at the point of tangency, it is perpendicular to the tangent.

3. If two chords are equal they intercept equal minor arcs.

4. If two minor arcs are congruent, the corresponding chords are congruent.

5. If two chords or secants in a circle are parallel, the arcs between them are equal.

$$\overline{AB} \parallel \overleftrightarrow{CD}$$

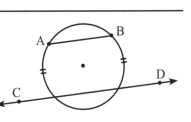

6. In a circle or in congruent circles, if two chords are equidistant from the center, they are congruent.

Given: $\odot H \cong \odot R$

$\overline{HT} \cong \overline{HS} \cong \overline{RU}$

$\therefore \overline{AB} \cong \overline{CD}$

Since $\odot H \cong \odot R$, and

$\overline{RU} \cong \overline{HS}$, $\overline{LE} \cong \overline{AB}$

and $\overline{LE} \cong \overline{CD}$.

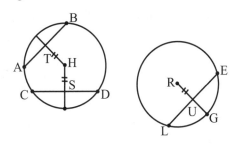

Circles

Common tangents of two circles: A tangent line can be tangent to two circles at the same time in various situations depending on the location of the circles in relation to each other. It is called a "common tangent." The diagrams below show the possible configurations of circles and common tangents. Even though a tangent is associated with two circles, all the same rules apply relative to each circle for tangent segment length, angle measure, perpendicularity to an intersecting radius or diameter, etc.

Two circles that are completely separate have four common tangents.

The tangents are numbered and the points of tangency are labeled with letters.

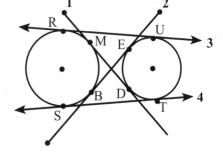

Two circles that are *tangent to each other externally* have three common tangents. Tangent externally means the circles share one common point and they are not inside each other.

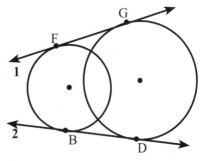

Two circles that intersect each other in two points have two common tangents.

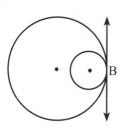 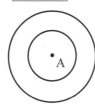 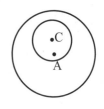

Two circles that are internally tangent have one common tangent line.	Concentric circles have no common tangents. They share the same center.	Circles that are inside each other but do not touch have no common tangents.

CIRCLE PROOF

Circle Proof – statement reason form

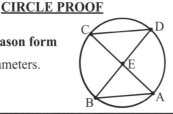

Given: Circle E with diameters.
$\overline{DB}$ and $\overline{CA}$
Prove : $\triangle CDE \cong \triangle ABE$

Statement	Reason
1. Circle E with diameters $\overline{BD}$ and $\overline{CA}$	1. Given.
2. $\angle CED \cong \angle AEB$	2. Vertical angles are congruent.
3. $\overset{\frown}{CD} \cong \overset{\frown}{AB}$	3. In a circle, two arcs intercepted by congruent central angles are congruent.
4. $\overline{CD} \cong \overline{AB}$	4. In a circle, if two minor arcs are congruent, their corresponding chords are congruent.
5. $\angle CDB \cong \angle CAB$	5. If two inscribed angles intercept the same or equal arcs, they are congruent.
6. $\triangle CDE \cong \triangle ABE$	6. AAS $\cong$ AAS

Note: This proof could be done several other ways.

- Segments EB, EA, CE, and DE are all radii and therefore congruent. Since $\angle CED$ and $\angle BEA$ are vertical angles, they are congruent. That makes arcs $\overset{\frown}{CD}$ and $\overset{\frown}{AB}$ congruent. In turn, that makes chords $\overline{CD}$ and $\overline{AB}$ congruent. The triangles are congruent by SSS $\cong$ SSS.

- Perhaps the easiest way to prove these two triangles are congruent would be to use SAS. The radii are all equal and $\angle CED \cong \angle BEA$. The triangles are congruent by SAS $\cong$ SAS. Any of these three methods is perfectly acceptable.

Circles

Example

Solve: In circle A, diameter $\overline{DE}$, secant $\overline{HG}$, and tangent $\overline{FG}$ are drawn. $\overline{DE}$ intersects $\overline{FG}$ at E. The ratio of the arcs $\overset{\frown}{HD} : \overset{\frown}{DB} : \overset{\frown}{BE}$ is 2:7:5. Find the measure of each of the numbered angles.

Solution: Let x be the ratio factor.

$\therefore m\overset{\frown}{HD} = 2x$; $m\overset{\frown}{DB} = 7x$; and
$m\overset{\frown}{BE} = 5x$. Since $\overline{DE}$ is a diameter,
$\overset{\frown}{DBE} = \overset{\frown}{DB} + \overset{\frown}{BE} = 180°$, also
$\overset{\frown}{DHE} = \overset{\frown}{DH} + \overset{\frown}{HE} = 180°$

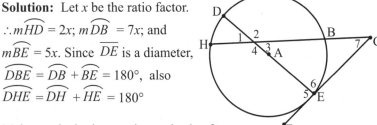

Make an algebraic equation and solve for x.

Find x

$\overset{\frown}{DB} + \overset{\frown}{BE} = 180°$

$7x + 5x = 180°$

$12x = 180°$

$x = 15°$

Find all the arc measures

$\overset{\frown}{DB} = 7(15) = 105°$

$\overset{\frown}{BE} = 5(15) = 75°$

$\overset{\frown}{DH} = 2(15) = 30°$

$\overset{\frown}{EH} = 180 - 30 = 150°$

Now work on the angles. It is very helpful to label the measure of each arc on your diagram. There are three different kinds of angles involved. Show the calculations.

1) Angles 1, 2, 3 and 4 are vertical angles formed by the intersection of a chord and a secant.

$m\angle 1 = (\frac{1}{2})(30 + 75) = 52.5$
$m\angle 3 = m\angle 1 = 52.5$
$m\angle 2 = (\frac{1}{2})(105 + 150) = 127.5$
$m\angle 4 = m\angle 2 = 127.5$

2) Angles 5 and 6 have vertices ON the circle. Since they are formed by a diameter intersecting with a tangent, both are right angles. Or you can use the formula $(\frac{1}{2})$ (arc intercepted).

$m\angle 5 = (\frac{1}{2})(150 + 30) = 90$
$m\angle 6 = (\frac{1}{2})(105 + 75) = 90$

3) Angle 7 is formed by a tangent and secant drawn from the same external point. It is an external angle and $= (\frac{1}{2})$(difference of the arcs it intercepts). $m\angle 7 = (\frac{1}{2})(150 - 75) = 37.5$.

Geometry Made Easy – Common Core Standards Edition

Circles

EQUATION OF A CIRCLE

Circle Equation using the Pythagorean Theorem: The general equation of a circle can be derived using the Pythagorean Theorem with a circle drawn on a coordinate axis.

Steps

Pythagorean Theorem: $a^2 + b^2 = c^2$

1) Draw a circle on a grid and label the center (h, k).

2) Draw a radius from the center to a point on the circle labeled with its coordinates, (x, y).

$(x - h)^2 + (y - k)^2 = r^2$

3) Draw two lines from (h, k) and from (x, y) that form right angles where they meet. Label them a and b.

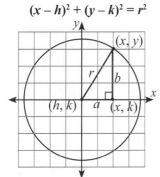

4) A right triangle is formed with a and b as its legs, and r as its hypotenuse.

5) Determine the lengths of a and b:
$a = (x - h)$ _and_ $b = (y - k)$

6) Substitute r for c in the Pythagorean Theorem and substitute the values found in step 5 for a and b. $a^2 + b^2 = c^2$

Examples

$(x - h)^2 + (y - k)^2 = r^2$

❶ Write the equation of a circle with its center at $(4, -3)$ and a radius of 6.

Solution: Substitute the coordinates of the center, $(4, -3)$, for h and k in the formula and substitute 6 for r.
$(x - 4)^2 + (y - (-3))^2 = 6^2$ which becomes $(x - 4)^2 + (y + 3)^2 = 36$
The circle equation is often left in the form shown but it can be multiplied out if necessary.

❷ Graph the following: $(x - 2)^2 + (y + 2)^2 = 25$

Solution: Determine the coordinates of the center (h, k) by comparing the equation with the known formula. h is 2 and k is -2. So the center is at $(2, -2)$. Plot that on the graph and label it as the center. Since 25 is the radius squared, then $r = 5$. Using the grid, place points on the graph 5 units vertically and horizontally from $(2, -2)$. Then estimate a few other points and sketch in the circle.

Points labeled: $(2, 3)$, $(-3, -2)$, $(2, -2)$, $(7, -2)$, $(2, -7)$, 5

Circles

Geometry Made Easy – Common Core Standards Edition

6.7

❸ Use the method of completing the square to find the center and radius of the circle given by the equation $x^2 + y^2 - 4x + 6y - 23 = 0$.

Plan: The given equation contains both x^2 and y^2 so completing the square will be required for both the x values and the y values. Rewrite the given equation in the general form of an equation for a circle. $(x - h)^2 + (y - k)^2 = r^2$ where the center of the circle is (h, k) and the radius is r.

Steps

1) Rewrite the equation so the x's and y's can be grouped.

$x^2 + y^2 - 4x + 6y - 23 = 0$

$x^2 - 4x + y^2 + 6y - 23 = 0$

2) Move the 23 to the right and collect the x's in one parenthesis, and the y's in another.

$(x^2 - 4x + \square) + (y^2 + 6y + \square) = 23$

3) Find the value needed, $\left(\dfrac{b}{2}\right)^2$, to complete the square within each parenthesis.

$\left(\dfrac{b}{2}\right)^2 = \left(\dfrac{-4}{2}\right)^2 = 4 \; ; \; \left(\dfrac{b}{2}\right)^2 = \left(\dfrac{6}{2}\right)^2 = 9$

4) Add this value to the x and y parentheses respectively and add both numbers to the 23.

$(x^2 - 4x + 4) + (y + 6y + 9) = 23 + 4 + 9$

5) Factor the x parenthesis and the y parenthesis.

$(x - 2)^2 + (y + 3)^2 = 36$

6) Compare the factors with the general form of the circle equation to find (h, k), the coordinates of the center. *Remember to use the opposite values of the numbers in the factors for h and k.*

$(x - h)^2 + (y - k)^2 = r^2$

$h = 2, \quad k = -3, \quad r^2 = 36$

Center: $(2, -3)$

7) Take the square root of the right side of the equation to find r, the radius.

Radius: 6

Circles

CIRCLES AND POLYGONS

CIRCLES RELATED TO TRIANGLES

Circumscribed Circle: A circle circumscribed about a polygon goes through all the vertices of the polygon. The polygon is completely inside the circle.

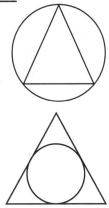

Inscribed Circle: A circle inscribed in a polygon is tangent to all the sides of the polygon. The circle is completely inside the polygon.

Concurrence – Diagrams and Theorems:

● **Incenter:** The point of concurrency of the three angle bisectors of a triangle. The incenter is inside the triangle.

Theorem: The incenter of a triangle is equidistant from each side of the triangle.

Incenter Diagram

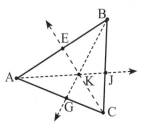

$\overline{AJ}$ bisects $\angle BAC$, $\angle BAJ \cong \angle CAJ$

$\overline{EC}$ bisects $\angle ACB$, $\angle ACE \cong \angle BCE$

$\overline{BG}$ bisects $\angle ABC$, $\angle ABG \cong \angle CBG$

K is the incenter of triangle ABC. The perpendicular distance from K to any side of the triangle is labeled r.

Inscribed Circle: A circle can be inscribed in a triangle by using the incenter as its center and the perpendicular distance to a side as the radius.

Inscribed Circle

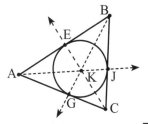

K is the incenter of $\triangle ABC$. Circle K is inscribed in $\triangle ABC$. The radius is the perpendicular distance from K to any side of $\triangle ABC$.

Circles

Circumscribed Circle: A circle can be circumscribed about a triangle by using the circumcenter as its center, and the distance from the circumcenter to a vertex as the radius.

- **Circumcenter:** The intersection of the perpendicular bisectors of the sides of a triangle. The intersection can be inside the triangle or outside.

Theorem: The vertices of a triangle are equidistant from the circumcenter.

Circumscribed Circle Diagrams

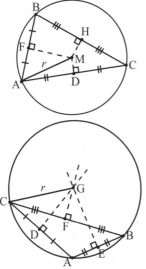

M is the circumcenter of $\triangle ABC$. Circle M is circumscribed about $\triangle ABC$. The radius is the distance from M to

any vertex of the triangle, A, B, or C.

G is the circumcenter of $\triangle ABC$. Circle G is circumscribed about $\triangle ABC$. The radius of G is the distance from G to any vertex of the triangle.

Note: Both the interior and exterior circumcenters are shown above.

REGULAR POLYGONS

Regular polygons can be *inscribed* (the circle is inside the polygon) in a circle. Circles can be *circumscribed* (the circle is outside of the polygon) about regular polygons and some that are not regular, but not all.

Inscribed Circle: Draw several diagonals of the polygon to locate the center. This point is the center of the inscribed circle as well. The radius of the circle is the perpendicular distance from the center to one of the sides of the polygon (also called the apothem). The inscribed circle will be tangent to each side of the polygon.

Example

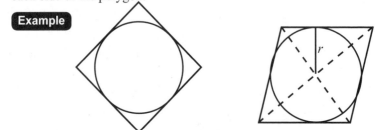

Circles

Circumscribed Circle: Find the center of the polygon using diagonals. The radius of the circumscribed circle is the distance from the center to a vertex of the polygon. A circumscribed circle goes through all the vertices of the polygon.

A circle can be circumscribed about a square or a rectangle, but not a rhombus!

This works

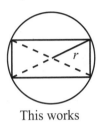

This works

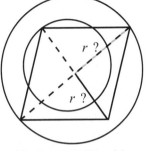

This does **NOT** work!

Example How can the properties of quadrilateral inscribed in a circle be demonstrated?

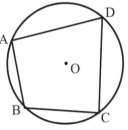

Given: Quadrilateral *ABCD* inscribed in Circle *O*.

Prove: $\angle A + \angle C = 180°$

Statement	Reason
1. $\overset{\frown}{BCD} = 2(\angle A)$ $\overset{\frown}{BAD} = 2(\angle C)$	An inscribed angle is equal to $\frac{1}{2}$ of the arc opposite it.
2. $\overset{\frown}{BCD} + \overset{\frown}{BAD} = 360°$	A circle contains 360°.
3. $2(\angle A) + 2(\angle C) = 360°$	Substitution.
4. $\angle A + \angle C = 180°$	Divide both sides of the equation by 2.

(The same procedure could be applied to angles *B* and *D* to show they are supplementary.)

Conclusion: When a quadrilateral is inscribed in a circle, opposite angles are supplementary.

Circles

CORRELATIONS TO COMMON CORE STATE STANDARDS

Common Core State Standards	Unit # . Section #

Congruence, Proof, and Construction (G.CO)

G.CO.1	2.3
G.CO.2	2.1
G.CO.3	2.2
G.CO.4	2.1
G.CO.5	2.2
G.CO.6	2.3
G.CO.7	2.3
G.CO.8	2.4
G.CO.9	1.2, 2.4
G.CO.10	1.2
G.CO.11	2.4
G.CO.12	2.5
G.CO.13	2.6

Similarity, Proof, and Trigonometry (G.SRT)

G.SRT.1	3.1
G.SRT.2	3.1
G.SRT.3	3.1
G.SRT.4	3.2
G.SRT.5	3.2
G.SRT.6	3.2
G.SRT.7	3.2, 3.3, 3.4
G.SRT.8	3.4

Extending To Three Dimensions (G.GMD)

G.GMD.1	4.1
G.GMD.2	4.2
G.GMD.3	4.3
G.GMD.4	4.2

Connecting Algebra and Geometry Through Coordinates (G.GPE)

G.GPE.1	4.1
G.GPE.2*	5.5
G.GPE.4	5.2
G.GPE.5	5.1
G.GPE.6	5.3
G.GPE.7	5.4

CORRELATIONS TO COMMON CORE STATE STANDARDS

Circles With and Without Coordinates (G.C)

G.C.1 .... 6.3

G.C.2 .... 6.1

G.C.5 .... 6.4

Modeling (G.MG)

G.MG.1   Throughout the units

G.MG.2

G.MG.3

* G.GPE.2 is not included in all of the content frameworks. It is included in the the Geometry Standards.

A

Abscissa 118
Adjacent Side 94
Analytic Proof 9
Angles 36, 46
    Central 141, 150, 155, 158
    Changing Degrees to Radians 154
    Changing Radians to Degrees 154
    Complementary 16, 100
    Exterior 16, 83, 159
    Finding 96
    Interior 16, 39, 83, 150, 152
    Trigonometric 93, 100
    Verticle 36
Arc 144
    Length 155
    Sector 155
    Subtend 144
    Sum 149
    Types 145
Area
    of a Circle 142
    of a Sector 158
Axiom 33

B

Betweenness 35
Bisector 14, 35

C

Cartesian Plane 118
Cavalieri's Principle 108
Center of Rotation 25
Chart Proof 8, 43, 44
Chord 146, 152, 160, 163
Circle 108, 139-171
    Area 142
    Changing Degrees to Radians 154
    Changing Radians to Degrees 154
    Chord 146, 152, 160, 163
    Circumcenter 170

Circumscribed 169-171
Concurrence 169
Constant Proportionality 153
Equation 167
Great 114
Incribed Angle 150
Inscribed Polygon 169-170
Incenter 169
Proof 163, 165-166
Regular Polygon 15,83,112,140,170
Secant 146, 161
Sector 155
Segments 160
Similarity 147
Tangent 146, 161, 164
Triangles 169-170
Circumcenter 170
Circumference 140
Circumscribed 169-171
Composite Transformation 18, 28, 63
Concurrence 169
Cones 109, 113
Congruent 13, 34
Constant of Dilation 58, 153
Construction 46
    Angle 47
    Bisector of an Angle 49
    Equilateral Triangle 53
    Equilateral Triangle in a Circle 56
    Line Segment 46
    Parallel Line 52
    Perpendicular Bisector 48
    Perpendicular Line 50, 51
    Regular Hexagon in Circle 55
    Square in a Circle 54
Coordinate Proof 9, 41, 42
Corollary 33
Cosine (Cos) 94, 96, 100, 106
Cross Section 111-114
Cube 110, 111
Cylinder 108, 113, 115

Index

D

Diameter 16, 144
Dilation 18, 58
Directrix 135-138
Distance Formula 9, 126, 132, 135

E

Equal vs. Congruent 34
Equation
    Circle 167
    Parallel Lines 123
    Perpendicular Lines 123
    Point-Slope 120
Euclidean Proof 4-6, 32
Exterior Angles 16, 65, 83, 159

F

Flow Proof 8, 82
Flowchart 8
Focus 135, 136
Formulas
    Distance 126
    Midpoint 126
    Slope 120

G

Geometric Definitions 13-16
Geometric Symbols 13-16
Geometric Terms 13-16
Glide Reflection 30
Great Circle 114

I

Incenter 169
Indirect Proof 4, 11
Inscribed Angle 150
Inscribed Circle 169-171
Interior Angles 16, 39, 83, 152
Invariant 18
Isosceles Triangle 26, 40, 67, 76, 115

L

Line Reflection 20
Line Segment 46
Line Symmetry 26-28
Lines 36
    Parallel 37, 122, 123
    Perpendicular 122-123

M

Magnitude 23
Mapped onto 26
Mapping 18
Mean Proportionals 77
Midpoint
    Definition 35
    Formula 126

N

n-gon 83-84
Non-Rigid Motion 18, 58

O

Opposite Side 94
Ordinate 118
Orientation 19
Origin 118

P

Parabola 135-138
Paragraph Proof 7
Parallel Line 122
    Equation 123
Parallelepiped 111
Parallelogram 43, 89
Perpendicular Lines 122
    Equation 123
Point Reflection 22
Point Symmetry 26-27
Polygon 83
    Inscribed 54, 140
    Perimeter 140
    Regular 15, 83, 170
    Similar 85

Index

Polyhedron 111
Postulate 4, 33, 35
Pre-Image 18
Prism 111
   Rectangular 112
   Right 112
   Triangular 112
Proof 4-6, 10, 32
   Analytic 9
   Circle 163, 165
   Contradiction 11
   Coordinate 9
   Euclidean 5-6
   Flow 8
   Flowchart 8
   Indirect 11
   Paragraph 7
   Rigid Motion 10
   Types of 4, 32
Properties of Equality 34, 111
Proportionality 153
Pyramid 109, 112
Pythagorean Theorem 67, 69, 75, 76, 91, 98, 99, 167
   Using Similarity 69
Pythagorean Triples 76

Q
Quadrilaterals 87

R
Radian Measure 153
Radius 144
Rectangular Prism 112
Rectangular Solid 108
Regular Polygon 15, 83, 170
Rhombus 27, 91
Right Prism 112
Right Triangle 67, 75
   Altitude 76
   Proportions 76-78
   Similarity 97
   Special 103, 105

Rigid Motion 10, 18
Rigid Motion Transformations 20
Rotation 24, 25, 115
   Center of 25
Rotational Symmetry 26

S
Secant 146, 161
Sector 144
   Area 158
Similar 13,
   Circles 147
   Polygons 85
Similarity 58, 70
Sine (Sin) 94
Slope 119
   Formula 120
Solid 111
Sphere 110, 114, 115
Subtend 144
Sum 149
Symmetry 26

T
(Tangent (Tan) 94
   of a Circle 146, 161, 164
Transformation of Functions 31
Transformational Geometry 18
Translation 22
Trapezoids 27, 88
Triangles 39
   Characteristics 64
   Circles 169-170
   Congruencies 79
   Isosceles 67, 76
   Right 67, 75
      Special 103, 105
   Similarity 70, 97
   Types 64
Triangular Prism 112

Geometry Made Easy – Common Core Standards Edition

Trigonometry 93
 Functions 94
  Inverse 96
 Geometric Problems 102
Two-Dimensional Objects 115
Types of Proof 32

V
Vector 18, 23
Vertex Form 135
 Derived 136

X
X-Intercept 119

Y
Y-Intercept 118

Geometry Made Easy – Common Core Standards Edition